288
Current Topi
in Microbiology
and Immunology

B.W.J. Mahy (Ed.)

Foot-and-Mouth Disease Virus

With 16 Figures, Mostly in Color

Springer

Professor Dr. Brian W.J. Mahy
National Center for Infectious Diseases
Centers for Disease Control and Prevention
Mailstop C 12, 1600 clifton road
Atlanta, GA 30333
USA

e-mail: bxm1@cdc.gov; virology@bellsouth.net

Cover illustration by Elizabeth E. Fry

ISSN 0070-217X
ISBN 978-3-642-06123-3 e-ISBN 978-3-540-27109-3

Springer is a part of Springer Science+Business Media
springeronline.com

Printed in Germany

Editor: Dr. Rolf Lange, Heidelberg
Desk editor: Anne Clauss, Heidelberg
Production editor: Andreas Gösling, Heidelberg
Cover design: design & production GmbH, Heidelberg

Printed on acid-free paper 27/3150/ag – 5 4 3 2 1 0

Preface

Foot-and-mouth disease (FMD) has been recognized in printed records dating from the sixteenth century, and since the eradication of rinderpest (cattle plague) in the early part of the twentieth century it has been recognized as the most important and feared disease of cattle and other domestic livestock. The beginning of the twenty-first century brought the worst outbreak of FMD ever experienced in England, which had been completely free of the disease for 33 years. This tragic epidemic, which spread to Northern Ireland, Scotland, France and the Netherlands with severe economic consequences, emphasized the need for further research into better methods for the detection and control of the disease.

FMD is caused by a small RNA virus which is highly contagious and can survive in meat and other animal products for long periods at normal pH levels. The virus typically infects cloven-hoofed animals, including cattle, goats, pigs and sheep, as well as a wide range of non-domesticated animals in regions of the world where FMD virus is endemic, such as the African continent.

There are seven recognized serotypes of FMD virus, with numerous subtypes, and as a consequence vaccine production and administration is complex and a major debate surrounds every disease outbreak regarding the relative merits of vaccination as opposed to the slaughter of all infected animals.

In 1999 a distinct strain of FMD type O virus was first detected in Kinmen, a Taiwanese island close to China, where it caused disease in cattle and mortality in goats in January 2000. In March 2000 this strain spread to Japan which had been free of FMD since 1908, and to Korea which had been free of FMD since 1934.

This strain, now known as the pan-Asian type O strain, also moved in a westerly direction to cause FMD outbreaks in the middle East and Europe, including the outbreak in England in 2001.

The seven chapters in this volume provide an account of the present knowledge and understanding of FMD pathogenesis and global epidemiology, the detailed structure of the virus itself and the properties of its RNA genome, the immune response of the host and the state of the art in vaccine production, and the nature of FMD virus evolution. It is clear that in all these areas there is still much more to learn about this fascinating virus. Because of its highly contagious nature, research work on FMD is restricted to a small number of laboratories worldwide that have adequate containment facilities. Despite this restriction, the recent progress in research on FMD which is described in this volume has provided a remarkable level of understanding of this unique virus.

October 2004, Atlanta, Georgia, USA
Brian W.J. Mahy

List of Contents

List of Contributors

(Their addresses can be found at the beginning of their respective chapters.)

CTMI (2005) 288:1–8

Introduction and History of Foot-and-Mouth Disease Virus

B. W. J. Mahy

National Center for Infectious Diseases,
Centers for Disease Control and Prevention, Atlanta, GA, USA
bxm1@cdc.gov

Abstract Foot-and-mouth disease (FMD) has been recognized as a significant epidemic disease threatening the cattle industry since the sixteenth century, and in the late nineteenth century it was shown by Loeffler and Frosch to be caused by a submicroscopic, filterable transmissible agent, smaller than any known bacteria. The agent causing FMD was thus the first virus of vertebrates to be discovered, soon after the discovery of tobacco mosaic virus of plants. It was not until 1920 that a convenient animal model for the study of FMD virus was established by Waldmann and Pape, using guinea-pigs, and with the later development of in vitro cell culture systems for the virus, the chemical and physical properties of FMD virus were elucidated during the remainder of the twentieth century, culminating in 1989 with a complete description of the three-dimensional structure of the virion. FMD virus is classified as a species in the *Aphthovirus* genus of the family *Picornaviridae*. The virus is acid labile, and the genome RNA contains a characteristic tract of polyC located about 360 nucleotides from the 5′ terminus. Seven main serotypes exist throughout the world, as well as numerous subtypes. The World Reference Laboratory for FMD is located at Pirbright, Surrey, UK and undertakes surveillance of FMD epidemics by serotyping as well as by genotyping isolates of the virus. A major epidemic of FMD occurred in the UK in 2001 and was caused by a virulent strain of FMD virus with origins in Asia. The advantages and some disadvantages of controlling FMD outbreaks by vaccination are discussed.

1
Introduction

Foot-and-mouth disease (FMD) is of great antiquity, and written records date back to a description of the disease by a monk, Hieronymous Fracastorius, who in 1546 described an epidemic which occurred in cattle near Verona, Italy. The disease became notorious as a perennial threat to the cattle industry over subsequent centuries, but it was not until the late nineteenth century that the pathogenic agent was discovered by two former pupils of Robert Koch, who were responding to a commission set up by the German government to discover the cause. Friedrich Loeffler and Paul Frosch worked originally in Greisswald but moved the project in 1909 to the island of Insel Rheims, in the Baltic Sea, where it was felt that work on FMD could be carried out without danger to livestock on the German mainland. The commission on FMD had been set up with the aim of producing a vaccine to prevent the disease, and Loeffler and Frosch took lymph fluid from vesicles caused by FMD and proceeded to filter the material through bacteria-proof filter candles in the hope that the infectious material would be retained, leaving behind an anti-toxin which could be used to confer passive immunity to healthy animals.

They were amazed when inoculated calves developed characteristic disease symptoms, and they were able to show that something had passed through the filter candles which was able to multiply in the infected animals. Loeffler and Frosch published their findings in four separate documents and a summary between 17 April 1897 and 12 August 1898, which were sent to the Minister of Culture. In the first report they stated, "Even the trial results show reliably that a bacterium which grows on a conventional substrate cannot be the etiological agent in FMD". The second report claimed that immunization against FMD was possible, and in the third report they concluded that the FMD agent "was small enough to pass through the pores of a filter which is impermeable to the tiniest known bacteria, so small, that even the best modern immersion system renders the agent unidentifiable under our microscope", which constituted the first description of a virus disease of animals (Loeffler and Frosch 1897, 1898). This was after D.I. Ivanovski had shown that the agent of tobacco mosaic disease would pass through a bacteria-proof filter candle,(Ivanovski 1892) but before Beijerinck developed the concept of a filterable virus he called contagium vivum fluidum (Beijerinck 1898).

For many years after its discovery, research on FMD virus was inhibited by the lack of a suitable experimental animal model to study the disease. Mice, rats and rabbits had been tried, but in 1920 Waldmann and Pape were able to show that guinea-pigs were susceptible by intradermal inoculation of the hind pad, so providing an important model that was used in many later studies of the virus.

2 The Virus

During the twentieth century, research on FMD virus led to a detailed understanding of its genetic and physical structure, culminating in a complete description of the three-dimensional structure of the virion by X-ray crystallography (Acharya et al. 1989).

The virus is classified as a member of the family *Picornaviridae*, which is comprised of viruses containing one molecule of single-stranded positive-sense genome RNA within icosahedral particles with no envelope, about 30 nm in diameter. Within this family, FMD virus is the type species of the genus *Aphthovirus*, the members of which are acid labile—FMD virus is unstable below pH 6.8. The genome RNA contains a variable length (100–400 nucleotides) polyC tract located about 360 nucleotides from the 5′ terminus. Besides FMD virus, the genus contains only one other virus, *equine rhinitis A*, which is related to FMD virus about 50% in nucleotide sequence across the entire genome.

3 Serotypes

In the early part of the twentieth century it became clear that FMD virus existed as more than one serological type, and initially two types were named, type O for Oise in France and type A for Allemagne (Germany) (Vallee and Carre1922) Later, type C was recognized as an additional type in Germany (Waldmann and Trautwein 1926). Some 30 years later, work at the Pirbright laboratory in England demonstrated three novel serotypes of the virus in samples which had been collected from FMD outbreaks in South Africa, and these were called SAT1, SAT2, and SAT3 (Brooksby 1958). The seventh and final serotype to be recognized, Asia

1, was present in a sample from Pakistan (Brooksby and Roger 1957), and extensive examination of further samples originating worldwide has not revealed the existence of other serotypes.

4 Genotypes

With the development of techniques for nucleotide sequence analysis, comparison of the nucleotide sequences of the capsid protein genes from many FMD viruses from different geographical sites of isolation showed that there was good correspondence with serotype differences, the sequences clustering into serotype-specific lineages upon phylogenetic analysis. The seven serotypes of FMDV cluster into distinct genetic lineages with approximately 30%–50% differences in the VP1 gene (Knowles and Samuel 2003). This differentiation can only be seen with the capsid region genes, and the sequences of other genes do not cluster in this way.

As a further aid to molecular epidemiology of FMDV, Samuel and Knowles have established "topotypes" of various FMDV serotypes, so, for example, FMDV type O can be divided into eight topotypes, each containing viruses which differ in VP1 sequence by at least 15% and are also geographically distinct (Samuel and Knowles 2001). With these techniques it has been possible to track the recent type O pandemic strain from Asia and into Africa and Europe.

5 The World Reference Laboratory

In 1924 the British Minister of Agriculture appointed a Foot and Mouth Disease Research Committee "to initiate, direct and conduct investigations into foot and mouth disease either in this country or elsewhere with a view to discovering means whereby the invasion of the new disease may be rendered less harmful to agriculture," and as a result The Animal Virus Research Institute was established in Pirbright in 1925, on the site of a tuberculosis quarantine station that was set up to ensure that pedigree cattle being exported to South Africa were free from TB. From 1925 until 1939 research was carried out at Pirbright under the auspices of the Foot and Mouth Disease Research Committee. After the outbreak of war, the full committee ceased to meet, but a smaller sub-committee held meetings. In 1947 the Foot and Mouth Disease Research Committee of the Ministry of Agriculture was re-appointed as an advisory body until 1957, when the Institute became an independent company, grant-aided by the Agricultural and Food Research Council (Mahy 1986). In 1958, the Institute was asked by the Food and Agriculture Organisation (FAO) to function as the World Reference Laboratory for Foot-and-Mouth Disease. Most recently, in 1988, the Institute lost its independence and became a component of the newly formed Institute of Animal Health, which operates on three geographically separated sites (Compton, Berkshire, Edinburgh, Scotland, and Pirbright, Surrey) under the auspices of the Biotechnology and Biological Sciences Research Council (BBSRC).

6
Transmission and Spread of FMDV

FMDV is highly contagious for cloven-hooved animals, and particularly so for cattle, but especially in Africa it can cause serious disease outbreaks in wildlife (Thomson et al. 2003). The virus has been reported to survive in infectious form for up to 12 years in soil attached to a Wellington boot and will survive for at least a year in cell culture medium held at 4°C. An extreme example of survival of the virus when spreading as an aerosol occurred in 1981, when virus from infected cattle in Brittany, France traveled as an aerosol to infect cattle in the Isle of Wight, a distance of over 250 km. Fortunately, the disease was detected in the UK cattle, and further spread of disease was prevented.

Another important means of spread of the disease is by carriage in milk or animal products such as frozen bone marrow and lymph nodes. Because of this, international trade laws ban the export of animal products from a country in which FMD is endemic, which may cause severe economic hardship to a country trading in meat and meat products. This is the main reason why FMD-free countries prefer to control epidemics by slaughtering infected animals and those in contact with them rather than relying on vaccination as a control strategy. There is evidence that vaccinated animals which are not completely protected may be a source of infection (Sutmoller et al. 2003). Although such animals may show no clinical signs, virus replication and shedding may occur.

7 Other Viruses Causing Vesicular Lesions

Because of the severe economic consequences of an outbreak of FMD, it has been important to distinguish it from other viruses causing vesicular lesions of FMD-susceptible animals. In 1932, an outbreak of vesicular disease in pigs occurred in Buena Park, CA, and as a result 19,000 pigs were killed and buried. However in the following year, a similar vesicular disease outbreak in San Diego, CA was examined more closely and found not to be due to FMDV. Instead, the causative virus was named vesicular exanthema of swine (VES) virus, and later found to be a calicivirus. Nevertheless, over the 10-year period beginning in 1939, numerous outbreaks of VES virus occurred throughout the United States, culminating in a national eradication program, with the last cases recorded in 1956.

Another disease of swine which may be confused with FMD is swine vesicular disease (SVD), which has been recorded throughout Europe and in Hong Kong and Japan. The clinical signs in pigs are very similar to those caused by FMDV, with lesions on the feet and snout. The SVD virus is an enterovirus that is closely related to human coxsackie virus B-5 (Knowles and McCauley 1997).

Finally, vesicular stomatitis virus, a rhabdovirus, causes a disease in horse, cattle and pigs which involves vesicular lesions in the mouth and on epithelia of the teats and feet. This disease occurs naturally in North and Central America and in the northern part of South America. The virus is transmitted mainly by the sand fly (*Lutzomyia shannoni*), but a number of other biting insects may also spread the disease. Differential diagnosis from FMD is difficult through clinical observation alone, except that horses are also affected during an outbreak, but examination of specimens in the laboratory is straightforward.

8 Vaccination

The first vaccine against FMDV relied on formalin-inactivated virus obtained from the tongue epithelium of infected cattle (Waldmann et al, 1937), but was eventually replaced by virus grown in vitro on bovine tongue epithelial cells (Frenkel 1951). By 1952 all cattle were vaccinated in the Netherlands by this procedure. This was extended to France and Germany over the next 10 years, but a much better method of FMDV production for vaccination purposes was developed at the Pirbright laboratory, when suspension cultures of BHK cells were found to support the growth of virus (Mowat and Chapman 1962; Capstick et al. 1965). Finally, problems with the inactivation of the virus by formalin (King et al. 1981) were overcome by the introduction of binary ethylene imine (BEI) as the inactivant (Bahnemann 1975).

Formulation of FMD vaccine has generally employed aqueous aluminum hydroxide and saponin as the adjuvant, but alternatively mineral oil emulsions are employed, and such vaccines are particularly used for vaccinating pigs, which do not respond well to vaccines based on aluminum hydroxide and saponin. With the use of such vaccines FMD was eventually eradicated from Europe, and vaccination ceased in 1991.

The main reason for stopping vaccination was based on the policy that vaccinated animals will not be accepted by countries which do not have the disease, and so their trading value is lost. However the dramatic UK epidemic of FMD in 2001 has alerted the world to the need for safe, effective vaccines against FMD virus should there be a recurrence of such an epidemic.

References

Acharya R, Fry E, Stuart D, Fox G, Rowlands D, Brown F (1989) The 3-dimensional structure of foot-and-mouth disease virus at 2.9-Å resolution. Nature 337, 709–716

Bahnemann HG (1975) Binary ethyleneimine as an inactivant for foot-and-mouth disease virus and its application for vaccine production. Arch. Virol. 47, 47–56

Beijerinck MW (1898) Ovewr een contagium vivum fluidum als oorzaak van de vlekziekte der tabaksbladen. Verst. gewone Vergad. Wis-en natuurk, Afd. K. Akad. Wet. Amst. 7, 229–235

Brooksby JB (1958) The virus of foot-and-mouth disease. Adv.Virus Res. 5, 1–37

Brooksby JB, Roger J (1957) In: Methods of Typing and Cultivation of Foot and Mouth Disease Viruses (Project 208 of OEEG). Paris, 31 pages

Capstick PB, Garland AJ, Chapman WG, Masters RC (1965) Production of foot-and-mouth disease virus antigen from BHK 21 clone 13 cells grown and infected in deep suspension cultures. Nature 205, 1135

Fracastorius H (1546) *De contagione et contagiosis morbis et curatione.* Bk.1, Chapter 12 (Venencia)

Frenkel HS (1951) Research on foot-and-mouth disease II. The cultivation of the virus on a practical scale in explantations of bovine tongue epithelium. Am. J. Vet. Res. 12, 187

Ivanowski DI (1892) On two diseases of tobacco. Sel'. Khoz. Lesov. 169, 108–121

King AMQ, Underwood BO, McCahon D, Newman JWI, Brown F (1981) Biochemical identification of viruses causing the 1981 outbreaks of foot-and-mouth disease in the UK. Nature 293, 479

Knowles NJ, McCauley JW (1997) Coxsackie virus B5 and the relationship to swine vesicular disease virus. In: The Coxsackie B Viruses, eds Tracy S., Chapman N.M. and Mahy B.W.J. Current Topics in Microbiology and Immunity 223, 153–167. Springer, Heidelberg Berlin New York

Knowles NJ, Samuel AR (2003) Molecular epidemiology of foot-and-mouth disease virus. Virus Res. 91, 65–80

Loeffler F, Frosch P (1897) Summarischer Bericht ueber der Ergebnisse der Untersuchungen zur Erforschung der Maul-und-Klauenseuche. Zent. Bakt Parasitkde Abt. I 22, 257–259

Loeffler F, Frosch P (1898) Report of the Commission for Research on foot-and-mouth disease. Zent. Bakt. Parasitkde. Abt.I 23, 371–391

Mahy BWJ (1986) Profile: The Animal Virus Research Institute, Pirbright, U.K. Microbiol. Sci. 3, 240–242

Mowat GN, Chapman WG (1962) Growth of foot-and-mouth disease virus in a fibroblastic cell line derived from hamster kidneys. Nature 194, 253–255

Samuel AR, Knowles NJ (2001) Foot-and-mouth disease type O viruses exhibit genetically and geographically distinct evolutionary lineages (topotypes). J. Gen. Virol. 82, 609–621

Sutmoller P, Barteling SS, Casas Olascoaga R, Sumption KJ (2003) Control and eradication of foot-and-mouth disease. Virus Res. 91, 101–144

Thomson GR, Vosloo W, Bastos ADS (2003) Foot and mouth disease in wildlife. Virus Res. 91, 145–161

Vallee H, Carre H (1922) Sur la pluralite des virus aphteux. C. R. Acad. Sci. Paris 174, 1498–1500

Waldmann O, Pape J (1920) Die kunstliche Ubertragung der Maul- und Klauensuche auf des Meerschweinchen. Berlin Tierarztl. Wschr. 36, 519–520

Waldmann O, Kobe K, Pyl G (1937) Die aktive Immunisierung des Rindes gegen Maul- und Klasuensuche mittels Formolimpfstoff. Zent. Bakt. Parasit. Infekt. 138, 401–412

Waldmann O, Trautwein K (1926) Experimentelle Untersuchungen uber die Pluralitat des Maul- und Klauenseuchevirus. Berlin Tierarztl.Wschr. 42, 569–571

CTMI (2005) 288:9–42

Foot-and-Mouth Disease: Host Range and Pathogenesis

S. Alexandersen[1, 2] (✉) · N. Mowat[3]

[1] Pirbright Laboratory, Institute for Animal Health, Ash Road, Pirbright, Woking, Surrey, GU24 ONF, UK
soren.alexandersen@bbsrc.ac.uk
[2] Danish Institute for Food and Veterinary Research, Department of Virology, Lindholm, DK-4771, Denmark
[3] 46 Boxgrove Avenue, Guilford, Surrey, GU1 1XQ, UK

Abstract In this chapter the host range of foot-and-mouth disease (FMD) under natural and experimental conditions is reviewed. The routes and sites of infection, incubation periods and clinical and pathological findings are described and highlighted in relation to progress in understanding the pathogenesis of FMD.

1 Introduction

Foot-and-mouth disease virus (FMDV) is classified within the *Aphthovirus* genus as a member of the *Picornaviridae* family (Bachrach

1968; Newman et al. 1973; King et al. 2000; Belsham 1993) and causes a severe vesicular disease, foot-and-mouth disease (FMD), of cloven-hoofed animals including domesticated ruminants and pigs and more than 70 wildlife species (Thomsen 1994).

Seven serotypes of FMDV, causing indistinguishable disease, have been identified, i.e. types O, A, C, Southern African Territories (SAT) 1, SAT 2, SAT 3 and Asia 1. Previous infection or vaccination with one serotype will not protect against subsequent infection with another, and within a serotype divergent strains of the virus may reduce the efficacy of existing vaccines (Kitching et al. 1989; Kitching 1998). The fact that FMDV has a wide host range, a high degree of contagiousness, a very rapid replication rate and multiple transmission routes and may cause a subclinical, persistent infection in ruminants makes FMD a difficult and expensive disease to control and eradicate. As a result, FMD is a major constraint to international trade in livestock and animal products.

FMD is endemic in large areas of Africa, Asia and South America, and the infection has a remarkable ability to spread over long distances and to cause epidemics in previously free areas, as seen, for example, in the 2001 epidemic in the UK, France and the Netherlands and in the outbreaks in South Korea and Japan in the year 2000 (Knowles et al. 2001b).

We summarise here the current knowledge of the host range and pathogenesis of FMD.

2 Host Range

With some minor exceptions, FMD affects members of the order Arteriodactyla, i.e. all cloven-hoofed animals including domestic and wild ruminants and pigs (Thomson 1994). When considering the host range of FMD it is important to distinguish between animals which (a) play a role in the natural epidemiology of the disease, (b) may play a role under certain conditions or cannot be excluded as being an epidemiological risk and (c) are susceptible to infection, and even may develop disease, under experimental conditions but appear to be without much, if any, importance under field conditions. The animals which under natural conditions are of greatest significance include cattle, pigs and small ruminants (sheep and goats) and, in particular in Asia and South America, the water buffalo. African buffalo play an important role as the natural maintenance host in Africa, but other wildlife such as impala and kudu may also be involved in the natural epidemiology of FMD. Animals which may contribute to the transmission of virus under certain conditions or which cannot be excluded as having some risk of transmission include deer, camels, llamas and alpacas, any animal of the order Arteriodactyla and Indian elephants. These animals may be of some significance if they get in close contact with livestock, for example when kept under farmed conditions or in zoos. Consequently, although these species do not appear to play an important role in the wild, they have to be considered as a potential risk, as mentioned above, in particular when they are kept under farmed or crowded conditions. As listed below, a large range of other animals may be infected but do not appear to be involved in the natural epidemiology of FMD. Finally, it should be mentioned that all animals, even highly resistant animals as for example horses and carnivores, can mechanically transfer the virus if they become contaminated and are subsequently in close contact with susceptible livestock.

2.1
Farm Livestock for Food Production

The animals of most significance in the natural epidemiology of FMD outside Africa are the species which are of major importance in the production of food, i.e. cattle, sheep, pigs, goats and other farmed cloven-hoofed animals. It is from this fact and the effects of the occurrence of the disease on the economies of those countries with highly developed agricultural industries that the major significance of FMD is derived.

Detailed descriptions of development of lesions and clinical signs are given in the paragraphs of this chapter devoted to describing the pathogenesis of the disease.

2.2
Free-Ranging Game Species

In countries with less well-developed animal production industries the existence of many species of cloven-hoofed animals has provided the possibility of reservoirs of infectious virus being established. These free-ranging species may on occasions come into close proximity, if not direct contact, with domesticated livestock, providing the opportunity for the transmission of virus and the initiation of an outbreak of the disease.

In the last 25 years there have been several publications reporting the incidence of FMD in wildlife as detected by either clinical signs or the presence of specific antibodies and the possibility or otherwise of this infection contributing to the epidemiology of the disease.

The most recent of these papers is a very comprehensive review of the subject (Thomson et al. 2003). In common with earlier reports it deals mainly with the infectiveness of the virus for the wildlife of Africa. Impala (*Aepyceros melampus*) have frequently been identified as having been involved in outbreaks of the disease in domestic livestock (Bastos et al. 2000; Thomson et al. 1984). African buffaloes (*Syncerus caffer*) have been recognised as major reservoirs of the SAT-type viruses and also may be carriers of the virus for several years (Anderson et al. 1979; Thomson et al. 1984; Condy et al. 1985; Hedger 1972).

In addition to the several species of deer [red, fallow, muntjac, sika and roe (Cottral and Bachrach 1968; Forman and Gibbs 1974; Forman et al. 1974; Gibbs et al. 1975a, b)], other species which have been reported to be susceptible to infection and may prove to be the source of virus in the origins of outbreaks affecting domestic farm livestock include white-tailed deer *(Odocoileus virginianus,* McVicar et al. 1974) and mountain gazelle (*Gazella gazella,* Shimshony et al. 1986). As mentioned above, there is little to indicate that deer play a significant role in the natural epidemiology of FMD under European conditions, although farmed deer may be considered a risk as they are kept closely together and possibly in contact with livestock; however, the possibility exists that free-ranging deer could play a role under certain circumstances and in other regions of the world and, for example, the mountain gazelle has been implicated in Israel (Shimshony et al. 1986). Consequently, when estimating the role of wildlife in FMD transmission and control, local circumstances clearly need to be taken into consideration. The water buffalo (*Bubalus bubalis*), which is widely used for agricultural operations throughout Asia and in South American countries, has also been reported as showing typical clinical disease after contact with infected cattle and subsequently transmitting the disease to goats. Outbreaks of FMD in cattle, sheep and pigs have been attributed to virus from infected water buffaloes (Dutta et al. 1983; Jerez et al. 1979; Moussa et al. 1979; Afzal et al. 1968).

The results of studies on the susceptibility of camels (*Camelus dromedarius*) to infection with FMDV reported by various groups are somewhat contradictory. The report of Farag and his colleagues (1998) stated that camels in contact with infected livestock showed no clinical signs of the disease and that 30 probang samples from camels were negative. Neither neutralising nor VIA antibodies were detected in these animals. In contrast, Moussa and his colleagues reported the isolation of FMDV from camels with an ulcerative disease syndrome (Moussa et al.

1987) and consequently camels may be a risk for transmission of FMD. Other members of the camelidae such as llamas and alpacas can be infected under experimental conditions; however, they appear to be relative resistant to infection and retain the virus for less than 14 days (Fondevila et al. 1995; Puntel et al. 1999).

The two species of elephant, i.e. the African (*Loxodonta africana*) and the Asian (*Elephas maximus*), appear to differ in their susceptibility to infection with FMDV. There is a report of the disease affecting African elephants in a circus in Italy (Piragino 1970), but in their natural habitat African elephants did not become infected when exposed to artificially infected cohorts or cattle (Howell et al. 1973). Also, there was no serological evidence for infection in elephants culled in the Kruger National Park during 30 years of investigations, and consequently the African elephant plays no role in the epidemiology of FMD (Bengis et al. 1984; Howell et al. 1973; Hedger et al. 1972).

In contrast, the Asian elephant is significantly more sensitive to infection (Hedger and Brooksby 1976) and the classical serotypes, O, A and C, together with the Asia 1 serotype have been recorded from outbreaks in India. Investigations into an outbreak of the disease in elephants used for ceremonial purposes in Nepal found that the titre of virus recovered from tongue epithelium tissue was of the order of $10^{6.0}$ to $10^{8.0}$ ID_{50} per gramme of tissue. The lesions in the mouth and feet of Indian elephants can be extensive and severe. This may be due to the invasiveness and virulence of the virus strains occurring in the Indian sub-continent, but another factor which might be important is that severity of the disease is to some extent related to the amount of mechanical stress to which the affected tissues are subjected. In the case of very large animals such as elephants, the stress on the feet due to the weight of the animal's carcase may result in the severity of some of the lesions described.

2.3 Investigations of Possible Vectors of the Disease

After the major epizootic of FMD in the UK in 1967–68 many investigations were made into the susceptibility of small British mammals to infection with the virus. This work had two main purposes, i.e. the identification of possible vectors of the virus resulting in a mechanism for the transmission of infection to farm livestock and also the possibility of identifying additional species useful in laboratory investigations.

Investigations were undertaken at the Animal Virus Research Institute, Pirbright into the susceptibility of small British mammals to infection with FMDV and shortly thereafter similar investigations were made with a number of Australian wild animal species. In experiments with small British wild mammals a range of susceptibility to the virus was demonstrated by Maureen Capel-Edwards (Capel-Edwards 1967, 1969, 1970, 1971a, b). This work can be summarised as follows:

1. *Myocastor coypus* (the coypu) was susceptible to FMD infection by inoculation and by contact with diseased cattle. The coypu may be considered in epidemiological surveys, although there is at present nothing to suggest that it has played any significant role in the natural epidemiology of FMD.
2. *Sciurus carolinensis* (the grey squirrel) was only of low-grade susceptibility, and it was unlikely that it would have any epidemiological significance.
3. *Arvicola amphibius amphibius* (the water vole) was very susceptible to FMD, the disease being lethal in the majority of these animals. There was also some evidence of sub-clinical infection in a contact animal. It is possible that these animals under certain conditions may play a part in the dissemination of the virus, although again, nothing suggests a significant role.
4. *Talpa europa* (the mole) was very susceptible to inoculation of the virus. However, the disease was uniformly and rapidly fatal, which would probably limit the epidemiological significance of this species.
5. *Rattus norvegicus* (the brown rat) was shown to be susceptible to virus inoculation and, moreover, continued to excrete the virus for several months. Contact rats and those fed with virus did not develop clinical signs but appeared to continue to excrete virus for several months; however, the pattern of detection of virus in the samples in the study appears to be somewhat erratic and may have been caused by particular circumstances in the experiments, e.g. the particular guinea-pig-adapted virus used in the experiments. The ability of this species to migrate considerable distances under pressure suggested that these animals could play a significant role in the epidemiology of the disease. Consequently, rodent control is considered to be an important component of efficient control of FMD in case of an epidemic. However, there is nothing to suggest that rodents can maintain the disease or are a significant risk of causing recrudescence when the infection is eradicated in domestic livestock.

Investigations at Pirbright with small wild Australian species showed that, although viraemia was detected in some of the wombats, possums, rat-kangaroos, bandicoots, water rats, echidnas and wild rabbits, infection was very mild in the small marsupials as clinical signs were very rare, but some of them did develop antibodies. However, a large species of marsupial, the tree kangaroo (*Dendrolagus matschiei*) did develop tongue lesions (Snowdon 1968). Consequently, among Australian wildlife there may be species constituting a substantial epidemiological risk if the disease was introduced; however, the actual risk cannot be predicted clearly on the basis of these experiments alone, as many species which are susceptible under experimental conditions, as mentioned above, often have no or only minor epidemiological significance under natural conditions in the field.

Apart from the species mentioned above, a large range of other species have been infected under experimental conditions. These include mice, guinea-pigs, rabbits and even cats, dogs, mink, monkeys, snakes, birds, chickens and embryonated eggs (Skinner 1954; Cottral and Bachrach 1968; Hyslop 1970; Sarhu and Dardiri 1979). However, in the case of, for example, mice, a useful laboratory model for FMD studies, the susceptibility is highly dependent on age (only very young mice being susceptible) and the genotype of the strain of mice (Skinner 1951, 1952, 1953; Bachrach 1968) and, moreover, the virus has to be directly injected into the animal and does not produce vesicles but rather infection and necrosis of the skeletal muscles (Platt 1956). Furthermore, infection of the species mentioned above, in particular, guinea-pigs (Henderson, 1949), rabbits and eggs, require multiple passage of the virus to adapt the virus to the new host. Consequently, infection of these species is possible but is not likely to play any role under field conditions as it requires forced passage of high doses leading to adaptation to the new species but often the virus is then less fit for the original host. The exact nature of adaptation is not known but is likely to require changes in the specificity of receptor binding as well as in specific virus-cell interactions. Sequence studies of mouse-, rabbit- and egg-passaged FMDV indicate that sequences in the 3A gene may be involved, and this gene has also been implicated in certain strains of FMDV with high virulence for pigs but low or no virulence for cattle (Dunn and Donaldson 1997; Beard and Mason 2000; Knowles et al. 2001a; Nick Knowles, personal communication).

Hedgehogs have long been given a special status in regard to susceptibility and possible risk in FMD transmission. This is because the hedgehog is highly susceptible to experimental infection, can transmit the infection to other hedgehogs or livestock and appears to excrete the virus in its breath when infected. Although there is a report describing lesions and isolation of FMDV from hedgehogs under field conditions

(McLauchlan and Henderson 1947), hedgehogs appear to play no or only a very minor role in the epidemiology of FMD. Nevertheless, during an epidemic it would probably be wise to exclude the access of hedgehogs to susceptible livestock.

2.4 Foot-and-Mouth Disease in Humans

There have been many references in the scientific literature in the past to the occurrence of FMD in humans. In the vast majority of these cases the appearance of blisters or vesicular lesions has been due to the effect of other agents—including viruses, e.g. some of the coxsackie A viruses (i.e. human enteroviruses in the *Picornaviridae* family) are capable of causing the condition described as "hand, foot and mouth disease". Nevertheless, there have been several reports of clinical signs of disease in humans after the ingestion of untreated milk from infected cattle, or resulting from coming in close contact with diseased animals. Clearly, to authenticate true infection of an unnatural species such as humans, the requirements of Koch's postulates have to be satisfied, which involves the isolation of the causal agent and its subsequent identification, its successful transmission to another known susceptible species and the demonstration of specific serological evidence in the original infected individual. Unfortunately, most of the reports of FMD in human lack this authentication and may not necessarily have been caused by FMDV infection.

One more recent case, in which FMDV has been properly identified (Armstrong et al. 1967), relates to the infection of an agricultural machinery salesman, who during the outbreaks of the disease which occurred in Northumberland in 1966 developed the clinical signs of the disease. He lived with his brother on a farm which was affected and consumed milk from one of the cows which later showed signs of the disease. Vesicular lesions developed in his mouth and hands and between some of his toes. Lesion material (epithelium) was collected, and during investigations at the Animal Virus Research Institute, Pirbright, virus of the serotype O was recovered and shown to be similar to that causing the outbreak on the farm. Subsequently it was also found that serum collected at 30 days after infection had high-titre type O antibody but was negative when tested against types A and C. There was therefore no doubt that this man had been infected by FMDV. In the UK 2001 FMD epidemic in which 2,030 outbreaks occurred among livestock, none of the 15 suspected cases in humans was positive for FMDV by reverse transcription-polymerase chain reaction (RT-PCR) analysis (Turbitt 2001).

As a conclusion, it has to be emphasised that infection of humans with FMDV is an extremely rare event, and where it does occur the results have only mild and transient consequences (Bauer 1997; Donaldson and Knowles 2001), and thus human infection does not appear to have any significant role in the natural epidemiology of FMD. However, people often play a significant role in passive transfer of the virus from infected animals or contaminated surfaces to susceptible animals, and may even passively carry the virus in the respiratory tract for a day or more (Sellers et al. 1970), and this is important to take into consideration in control programmes.

3 Clinical Signs of the Disease and Pathogenesis

Under field conditions susceptible animals may be infected by FMDV as a result of direct or indirect contact with infected animals or an infected environment. When animals are close together the transfer of airborne droplets and droplet nuclei (aerosols) from the breath of infected animals to the respiratory tract of recipient animals is probably the most common form of transmission, although the virus may also gain entry into a susceptible host through damaged integument. Long-range airborne transmission of virus is an uncommon, but under certain conditions important, route of infection and requires the chance combination of particular factors, including the animal species, the number of the excreting and inhaling animals and favourable topographical and meteorological conditions.

3.1 Routes of Infection

Most of our understanding of the routes of infection derives from experimental studies in animals infected with FMDV by simulated natural methods (by direct or indirect contact with infected animals, contaminated products or fomites or by exposure to virus aerosols from infected animals) or in animals infected by artificial methods, including direct injection of virus. From these studies the minimum infective doses for the main livestock species infected by different routes can be estimated. Such values are not absolute but only indicative, as the statistical significance of the estimates is less than optimal because of the practical constraints on such studies (a limited number of animals, virus strains and other variables studied) and because the various methods used for estimating the dose cannot be directly compared.

The most common and efficient mechanism of spread of FMD is by direct contact, which, as mentioned above, may be initiated by the deposition of droplets or droplet nuclei (aerosols) in the respiratory tract or by mechanical transfer of virus from infected to susceptible animals and subsequent virus entry through cuts or abrasions in the skin or mucosae. The intact cornified (keratinised) epidermis provides good protection against virus entry, but pre-existing damage to the intact integument predisposes to direct infection. Consequently contact transmission, especially in pigs, which are quite resistant to aerosol inhalation (Alexandersen and Donaldson 2002; Alexandersen et al. 2002a), may be enhanced by physical contact with infected excretions or secretions containing large amounts of FMDV. Transmission of virus may also occur indirectly, via any contaminated surface or product, i.e. contaminated personnel, vehicles and all fomites. Farming or disease control activities, such as shearing or de-worming or clinical examination and blood sampling procedures commonly done during FMD epidemics, greatly increase the risk of indirect spread of the virus by increasing the contact rates and by facilitating virus entry through traumatized epidermis.

Direct entry of the virus into superficial epithelia or into the body through pre-existing damage is often simulated in experimental studies. Transmission by this route has occasionally been incriminated in iatrogenic disease due to the use of contaminated instruments (e.g. needles) or medicines, e.g. injection of FMDV-contaminated pituitary extract and FMD vaccines containing live virus when sub-optimal inactivants were used (Beck and Strohmaier 1987). Intradermal/subdermal injection of virus into the tongue, coronary bands or heel bulbs in cattle, sheep and pigs, respectively, or application of a suspension of virus to damaged (scarified) skin, targets the highly susceptible epithelial regions (Henderson 1949, 1952; Burrows 1966b, 1968b) and simulates natural infection through damaged skin by overcoming the normal protective effect of the intact integument. The dose to establish infection by this route may be as low as 100 tissue culture 50% infective doses ($TCID_{50}$) or less (Sellers 1971); in comparison a single infected animal may excrete as much as 10^{14} $TCID_{50}$ in a single day (Alexandersen et al. 2003c). Experimental exposure of the surface of the tongue of cattle to more than 10^7 infectious units for 10 min only led to FMD if the cornified epithelium of the tongue was first damaged by scratching with a needle (Cottral et al. 1965), indicating, as already discussed, that direct entry into the epithelia may be an important route of natural transmission in animals with pre-existing lesions due to trauma or intercurrent disease. Direct entry of the virus into the circulation, as simulated by intravenous inoculation, also results in infection, but this method appears less efficient and more variable than the routes described above targeting epithelia (Henderson 1952). Intramuscular inoculation is relatively inefficient and requires a dose of 10^4 $TCID_{50}$ or more (Burrows et al. 1981; Donaldson et al. 1984).

Several recent outbreaks of FMD have been linked to the entry of virus in contaminated material which has subsequently been fed to animals. For example, the South Africa 2000 and UK 2001 epidemics have been ascribed to the feeding of unheated waste food to pigs (Knowles et al. 2001b; Alexandersen et al. 2003a). Experimental infections by the oral route have indicated that the dose for pigs and ruminants is 10^4–10^5 and 10^5–10^6 $TCID_{50}$, respectively (Sellers 1971), but it is conceivable that abrasions or other damage to the epithelium of the mouth, for example caused by pointed bits and pieces of e.g. bone commonly present in waste food, may facilitate oral infection, reduce the dose required to infect individual animals and significantly increase the risk of infection by contaminated waste food.

Under specific epidemiological, climatic and meteorological conditions, short-distance aerosol transmission, which as mentioned above is a highly efficient route of infection of ruminants, may be extended to airborne transmission over a significant distance. This is particularly a risk when large numbers of pigs are infected because pigs excrete large quantities of airborne virus (up to $10^{5.6}$–$10^{8.6}$ $TCID_{50}$ per pig per day), whereas ruminants excrete less virus in their breath (10^4–10^5 $TCID_{50}$ per day) but, in contrast to pigs, are highly susceptible to infection by inhaled virus (Sellers and Parker 1969; Donaldson et al. 1970, 1982a; Donaldson and Ferris 1980; Alexandersen and Donaldson 2002; Alexandersen et al. 2002a). It has been established that ruminants can be infected experimentally by airborne exposure to only 10 $TCID_{50}$, whereas to infect pigs by this route more than 10^3 $TCID_{50}$ are required and infection only occurs if virus is delivered at a high concentration (Donaldson et al. 1970; Donaldson and Ferris 1980; Donaldson 1986; Gibson and Donaldson 1986; Donaldson et al. 1987; Donaldson and Alexandersen 2001; Alexandersen and Donaldson 2002; Alexandersen et al. 2002a). Therefore, most often the pattern of airborne spread of FMD is from pigs to cattle and sheep downwind. Because cattle are larger than sheep, they inhale more air in a given time and are therefore likely to be more readily infected than sheep by the airborne route. FMDV isolates vary in the amount of virus released in the breath of infected animals, and consequently the distance over which airborne spread can occur is likely to be dependent on the isolate of FMDV involved in a specific outbreak. On the current knowledge and computer simulation, it has been estimated that some isolates, for example, the type O UK 2001 strain, are unlikely to be spread more than up to 20 km by the wind even when the meteorological conditions are optimal for spread and many infected pigs provide the source (Donaldson et al. 2001; Alexandersen and Donaldson 2002; Donaldson and Alexandersen 2002). However, some isolates, e.g. FMDV C Noville, may have the potential for spread up to 300 km by the wind (Gloster et al. 1981, 1982; Donaldson et al. 1982a,b; Sorensen et al. 2000, 2001). Long-distance airborne spread is only likely to occur when the infectivity and concentration of virus in the plume are relatively stable because of the specific climatic and topographical conditions, i.e. a relative humidity above 55% and minimal mixing of the air. Mixing of air may occur from the turbulence following passage of wind over hills, trees or buildings. Conditions favourable to transmission occur under a continuous steady or slight wind, cloud cover and a level topography such as the passage of the plume over, e.g. large tracts of water (Donaldson et al. 2001; Alexandersen and Donaldson 2002; Donaldson and Alexandersen 2002). Aerosols from infected animals contain large, medium and small particles excreted as droplets and droplet nuclei in the breath. For FMDV-infected animals it is a significant characteristic that a large proportion (30%–65%) of the excreted airborne infectivity is associated with small to medium-sized particles (<6-μm diameter). When inhaled by recipient animals, large particles will be deposited mainly in the nares whereas medium-sized and small particles will be deposited in the pharynx, trachea and bronchi and in the small bronchioles and bronchiolar-alveolar junction, respectively (Hatch and Gross 1964; Sellers and Parker 1969; Donaldson et al. 1970, 1987; Donaldson and Ferris 1980; Alexandersen, unpublished data). Larger droplets will tend to sediment rapidly, and although turbulence may keep them suspended longer, such turbulence will also significantly dilute the effective concentration of virus by mixing and consequently large droplets mainly play a role for short-distance spread. However, particles of less than 6-μm diameter will not be affected much by gravity and therefore can be transported over long distances (Gloster et al. 1981) and, as described above, are most likely to be deposited in the upper and middle regions of the respiratory tract. Particles landing in the nares as well as in trachea and bronchi will be taken towards the pharynx by the muco-ciliary escalator and thereby concentrate the virus in the pharynx, a site of entry and early replication of FMDV. Thus, in particular, ruminants are highly susceptible to aerosol infection because the virus is concentrated in the pharyngeal region, which is highly susceptible to infection. If FMDV is given as a liquid inoculum directly into the nares instead of as a natural aerosol, ruminants may still be infected but the dose is much larger than that for aerosol infection, i.e. around 10^4–10^5 $TCID_{50}$ (McVicar and Sutmoller 1976). As for the airborne route of infection, the site of virus entry is probably the pharynx (McVicar and Sutmoller 1976); however, only a fraction of the dose is likely to reach the pharynx, which may explain the relatively high dose needed by this route although, as mentioned above, it is possible that infection by this route could also be facilitated by pre-existing epithelial damage.

3.2 Primary and Secondary Sites of Infection

As mentioned above, the pharyngeal area is the usual primary site of infection except when the virus directly enters into the cornified epithelia or the circulation by damage to the intact integument (Garland 1974; McVicar and Sutmoller 1976; Burrows et al. 1981). In contact- or aerosol-exposed animals, virus may be demonstrated in the pharynx for 1 to 3 days before a viraemia or clinical disease can be detected (Burrows 1968a; McVicar and Sutmoller 1976; Burrows et al. 1981; Alexandersen et al. 2002b and c; Zhang and Alexandersen, unpublished data; Garland, unpublished data). The dorsal surface of the soft palate and the adjacent nasopharynx are sites of particular significance for initial virus entry and replication as demonstrated originally by probang sampling and subsequently by in situ hybridisation and "real-time" RT-PCR (Murphy et al. 1999; Zhang and Kitching 2000, 2001; Alexandersen et al. 2001; Oleksiewicz et al. 2001; Zhang and Alexandersen, unpublished data). The tonsillar area may also play a significant role in the initial infection, in particular in sheep (Burrows 1968b), probably because the epithelium covering part of the tonsil may also be of a transitional type and because in sheep the tonsils are located immediately adjacent to the dorsal soft palate, resulting in physical contact.

The epithelial cells in the pharyngeal region play a special role in primary infection. Most of the oral cavity is covered by cornified/keratinised (i.e. having a layer of dead cells at the surface) stratified squamous epithelia, whereas the anatomical regions mentioned above, i.e. the dorsal soft palate, the roof of the pharynx and part of the tonsil, are covered by a special non-cornified, stratified squamous epithelia and therefore, in contrast to intact cornified epithelia, have live cells exposed on the surface and consequently may allow easy access, and, provided the right receptors are present, efficient virus entry. FMDV entry into cells in vivo is believed to involve attachment of the RGD loop of VP1 on the viral capsid to host integrins such as $\alpha v\beta 6$, $\alpha v\beta 3$, $\alpha v\beta 5$ or $\alpha v\beta 1$ on the surface of target cells (McKenna et al. 1995; Rieder et al. 1994, 1996; Jackson et al. 1997, 2000a and b, 2002; Sa-Carvalho et al. 1997; Neff et al. 1998, 2000). Little is known about the relevance of the FMDV receptors in relation to host range, target cells or persistence. Although it may appear likely that the receptor(s) would be an important determinant of host range, the above-mentioned studies have in several instances used human or non-livestock genes apparently allowing efficient entry and replication of FMDV. Consequently, it may be possible that other host factors are important for efficient replication of FMDV in vivo and that the presence of appropriate receptors on a cell is in itself not sufficient to allow replication of FMDV.

After initial replication in the pharynx, or in the skin if the virus has entered directly through damaged integument, virus is spread through regional lymph nodes (Henderson 1948) and into the circulation (Burrows 1968a; McVicar and Sutmoller 1976; Burrows et al. 1981; Alexandersen et al. 2002b and c; Zhang and Alexandersen, unpublished data; Garland, unpublished data). This can be detected as a plasma/serum-associated viraemia usually lasting for 4–5 days (Cottral and Bachrach 1968; Alexandersen et al. 2002c; Alexandersen et al. 2003b; Garland, unpublished data), resulting in seeding of secondary sites and multiple cycles of viral replication and spread, in particular in the cornified epithelia of skin, tongue and mouth where the main viral amplification occurs (Oleksiewicz et al. 2001; Alexandersen et al. 2001; Hess et al. 1967; Burrows et al. 1981; Zhang and Alexandersen, unpublished data). Although vesicular epithelia clearly contain the highest concentration of virus, apparently normal skin, both hairy and hairless, also contains significant amounts (Alexandersen et al. 2001). Experimental studies suggest that lymph nodes as well as lymphocytes and macrophages (including alveolar macrophages) play little or no part in FMDV replication and that any virus present in lymphoid organs is produced elsewhere, i.e. the epithelia of the pharynx, mouth and skin (Cottral et al. 1963; Burrows et al. 1981; Alexandersen et al. unpublished data).

3.3 Virus Clearance or Persistence

The host reaction, in particular antibody production, can be detected from 3–4 days after the first clinical signs and is usually sufficient to clear the virus, except in carrier ruminants which develop a persistent infection of the pharyngeal region. Immunity to FMDV is primarily mediated by circulating antibodies which are highly efficient in clearing virus from the circulation. However, clearance of virus from surfaces such as the nasal and oral surfaces is less efficient and, similarly, virus may remain at detectable levels in epithelium for up to 10–14 days (Oliver et al. 1988). Clearance from the oesophageal-pharyngeal (OP) region is even less efficient, even in animals that do not develop the carrier state, and although recovery from infection and protection from disease by inactivated vaccines or passively transferred antibodies are correlated with the concentration of circulating antibodies, these antibodies do not efficiently protect against local pharyngeal infection and ruminants may develop a persistent infection at this site (Brown and Cartwright 1960; Hess et al. 1967; McVicar and Sutmoller 1969b, 1974, 1976; Francis and Black 1983; Francis et al. 1983; Black et al. 1984; Hamblin et al. 1987; McCullough et al. 1992; Aggarwal et al. 2002; Alexandersen et al. 2002b). This concept of FMDV carrier animals was initially based on field experience but was confirmed when van Bekkum et al. (1959a and b) showed infectious virus in the "saliva" (OP fluid) in some recovered cattle for many weeks after infection. A carrier of FMDV is defined as an animal from which virus can be detected for at least 28 days after infection (Sutmoller and Gaggero 1965; Burrows 1966a; Sutmoller et al. 1968). Carriers have been found in a proportion of infected cattle, sheep and goats, whereas pigs appear to efficiently clear FMDV infection in 3–4 weeks or less and so do not become carriers (see Salt 1998; Alexandersen et al. 2002b).

A proportion of ruminant animals exposed to FMDV become carriers, irrespective of whether they are fully susceptible or protected from disease as a result of vaccination or recovery from infection; the percentage of animals which become carriers under experimental conditions is variable but averages 50%. The infectivity titre of virus in OP samples from carriers is usually low, virus recovery is intermittent and the titre declines over time. The animal species and the specific strain of FMDV are determinants in the development and duration of the carrier state. The maximum reported duration of the carrier state in different species is as follows: cattle, 3.5 years; sheep, 9 months; goat, 4 months; African buffalo, 5 years (Hedger 1972; Thomson et al. 1984; Condy et al. 1985; Hedger and Condy 1985; Alexandersen et al. 2002b). Other cloven-hoofed wildlife species, including deer and impala, which may become acutely infected, do not become carriers or only do so for a relatively short period and are therefore unlikely to play an important epidemiological role as carriers (McVicar and Sutmoller 1969a; Hedger et al. 1972; McVicar et al. 1974; Gibbs et al. 1975a, b; Thomson et al. 1984; Bastos et al. 2000). Information on the water buffalo is very limited; a single experimental study from Egypt suggested a carrier state lasting at least 6 weeks (Moussa et al. 1979). Naturally, the presence of a symptomless FMDV carrier state has contributed significantly to the severe trade implications of FMD.

3.4 Incubation Periods

The incubation period for FMD depends to a high degree on the dose of virus received and on also the route of transmission, the specific strain of FMDV and the animal species and husbandry conditions and is therefore highly variable (Alexandersen et al. 2003a, b, c). The range of incubation periods for farm-to-farm spread by indirect contact is normally 4–14 days, and this is also taken as the expected range for farm-to-farm airborne spread (Sellers and Forman 1973). For farm-to-farm spread resulting from direct contact with an infected animal the incubation period may range from 2 to 14 days, and this is also often the case for on-farm spread (Garland and Donaldson 1990), although it typically is 2–6 days and may be only 1 day, especially under high-challenge conditions and depending on the degree of contact (Alexandersen et al. 2002c; Alexandersen et al. 2003b; Garland, unpublished data; Alexandersen et al. unpublished data). Under field conditions the dose and intensity of FMDV contact are influenced by a number of factors, including in particular the stocking density, animals being outside or housed and, if housed, the extent of ventilation. Handling of the animals, especially in and around the mouth and nares as associated with examination by veterinarians, when farmers or livestock dealers examine animals on farms or at markets or when gathering of livestock for shearing, dipping, deworming, vaccination, transport, marketing, etc. will also accelerate the rate of spread of virus on an infected premises and shorten the incubation period.

3.5 Clinical Signs

This section describes the clinical signs of FMD; however, it should be emphasised that the clinical diagnosis of FMD may sometimes be difficult, for example, in small ruminants in which clinical signs are often mild (Callens et al. 1998; Barnett and Cox 1999; Donaldson and Sellers 2000; Alexandersen et al. 2002c; Hughes et al. 2002). In addition, particular strains of FMDV may be of low virulence for some species and highly virulent for others (Donaldson 1998) and other viral vesicular diseases, such as swine vesicular disease, vesicular stomatitis and vesivirus infection, cannot be easily distinguished from FMD solely on the basis of clinical findings. Consequently, a definitive diagnosis of FMD requires laboratory confirmation.

FMD is an acute febrile disease characterised by the formation of vesicles in and around the mouth and on the feet. Lameness and inappetence are also often characteristic features. Lesions may initially be observed as blanched areas, which subsequently develop into vesicles at sites of local irritation or abrasion. Consequently, lesions are most often observed in and around the mouth and on the feet but may also be seen on the snout or muzzle, teats, mammary gland, prepuce, vulva and other sites of the skin and mucosae. Animals kept outside on soft ground or inside on soft bedding are less likely to develop severe foot lesions and show obvious lameness. Clinical disease is usually severe in pigs, characterised by severe vesicular lesions affecting the coronary band, the bulb of the heel and the interdigital space (Fig. 1) but sometimes also including the dorsal and rostral surface of the snout, accessory digits and pressure points on knees, hocks and elsewhere (Kitching and Alexandersen 2002). The clinical signs in cattle are most often obvious and include the drooling of saliva and rather severe vesicular mouth lesions (Fig. 2); however, lesions may also be seen on the feet (interdigital space, bulb of the heel and the coronary band) and elsewhere. In sheep and goats the signs are usually rather mild and tend to be characterised by superficial lesions (Fig. 3) that heal rapidly (Donaldson and Sellers 2000). Lesions in the mouth of large and, to a lesser extent, small ruminants are most often seen on the dental pad and the dorsum of the tongue but may also be seen on the lips, gums and cheeks and sometimes on the hard palate. In pigs, lesions in the mouth are not always a consistent finding and when present such lesions are most often located either far back on the dorsum or as tiny lesions at the tip of the tongue. The rupture of vesicles, especially on the feet or teats, may predispose the affected areas to

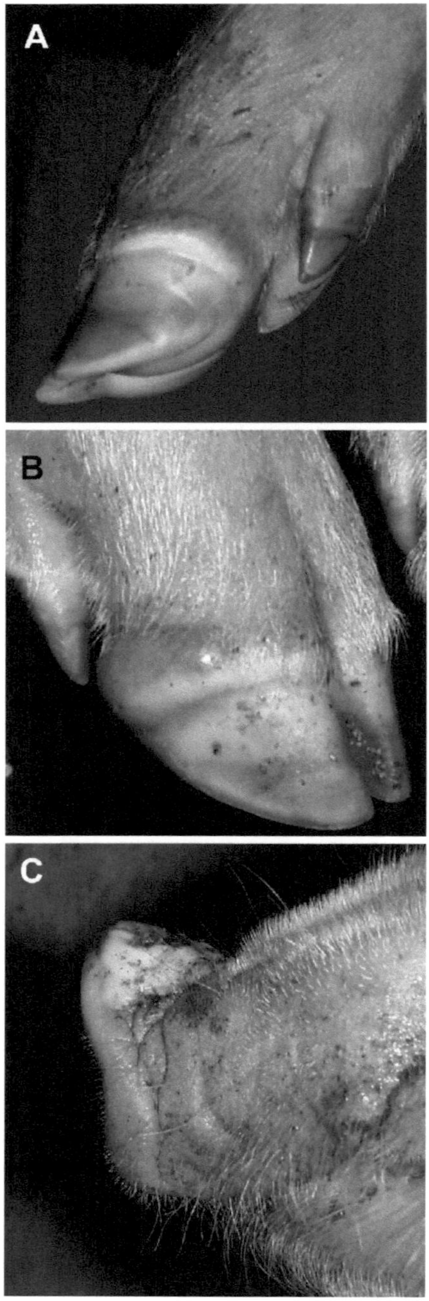
A
B
C

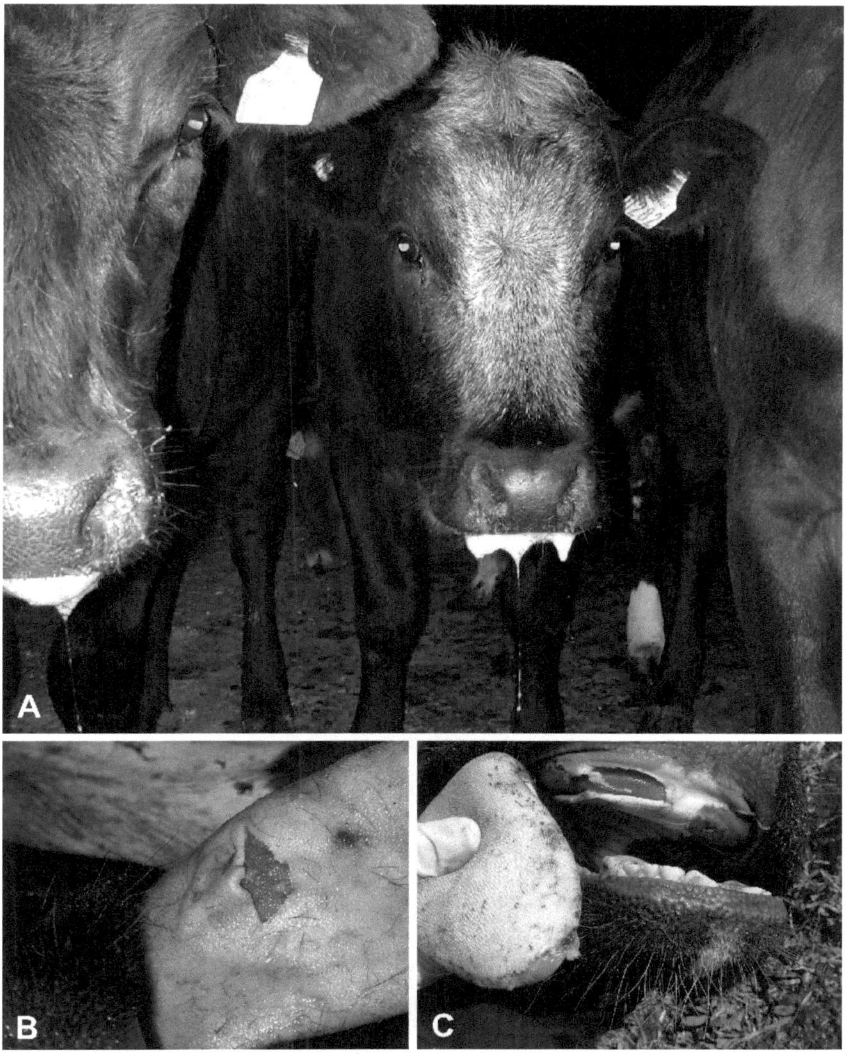

Fig. 2A–C FMD in cattle. **A** Excess salivation (drooling) is obvious. **B** Just ruptured vesicle on the tongue. **C** Severe erosions on the dental pad and gums. The pictures were taken in the field during the UK 2001 type O FMD epidemic and are courtesy of DEFRA, Animal Health Office, Carlisle, UK

Fig. 1 A–C FMD generalized lesions in pigs 3–5 days after exposure to pigs inoculated with FMDV O_1 Lausanne. Unruptured vesicles are evident along the coronary bands. C Lesion on the snout of a pig inoculated with FMDV O UK 2001

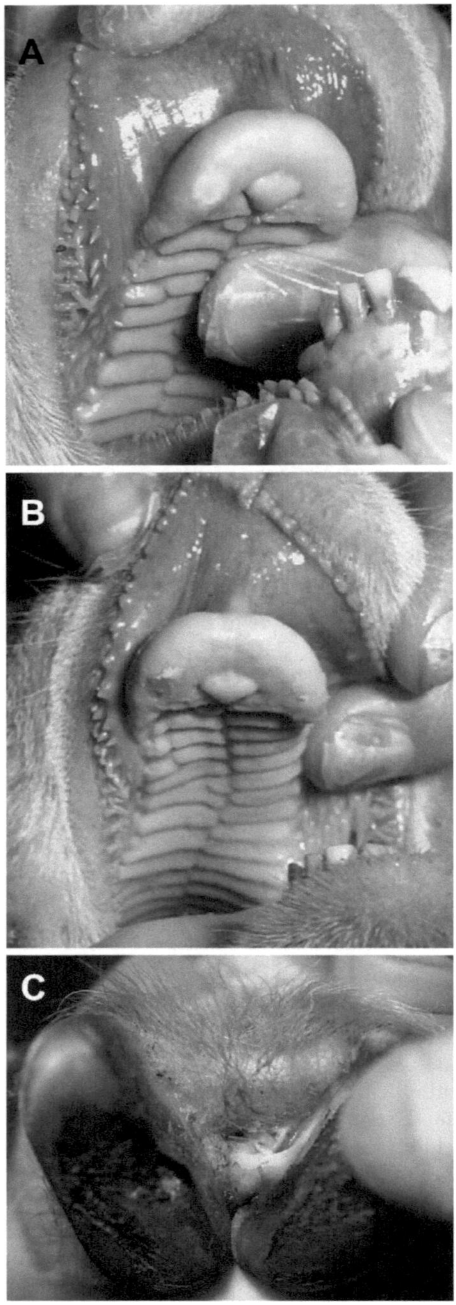
A
B
C

secondary infections, but lameness following rupture of foot lesions is less obvious. Virus strains may differ in their virulence for different species. For example, the O Taiwan 1997 strain caused severe lesions in pigs, but no cases were seen in ruminants (Dunn and Donaldson 1997). The marker for the severe virulence of the O Taiwan 1997 strain for pigs and the absence of virulence for cattle is associated with changes in the 3A gene of the virus (Knowles et al. 2001a). Virulence of FMDV strains may also vary between breeds of animal and sometimes within a breed, probably because of genetic or physiological factors.

3.6 Pathology

Very early lesions are only detectable by microscopical examination (Gailiunas 1968; Yilma 1980), and it is characteristic that even apparently normal skin with no macroscopical or histopathological changes may contain significant amounts of virus (Alexandersen et al. 2001). The first histopathological changes can be observed in the cornified, stratified squamous epithelium and are characterised by ballooning degeneration and increased cytoplasmic, eosinophilic staining of the cells in the stratum spinosum and the onset of intercellular oedema within the dermis (Fig. 4). This is followed by necrosis and subsequent mononuclear cell and granulocyte infiltration; the lesions, now macroscopically visible, develop into vesicles by separation of the epithelium from the underlying tissue and filling of the cavity with vesicular fluid (Fig. 4). In some cases vesicular fluid production may be high and the vesicles large; in other cases the amount of fluid may be limited and the epithelium undergoes necrosis or is torn off by physical trauma without the formation of a conspicuous vesicle. This variability is likely due to a combination of the virulence of the specific strain of FMDV, the thickness of the skin affected and the husbandry conditions as these affect the physical stress on the skin. It should also be noted that ruptured lesions are sometimes seen on the pillars of the rumen.

Fig. 3 A–C FMD generalized lesions in sheep infected with FMDV O_1 BFS 1860. **A** An unruptured vesicle can be observed on the dental pad; however, such lesions rupture early (**B**), leaving shallow erosions which heal within a few days. **C** Ruptured vesicle in the interdigital area

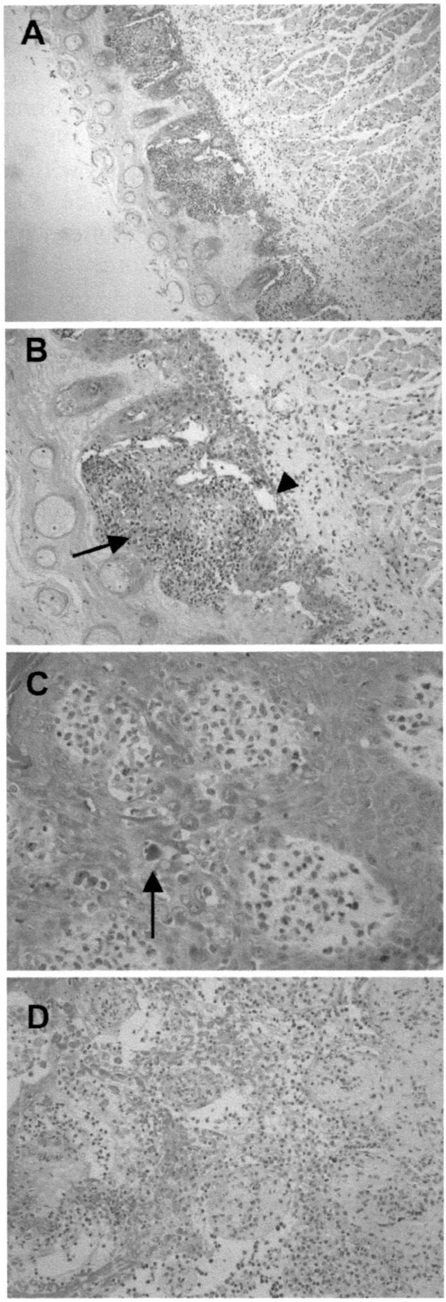
A
B
C
D

Mortality in young animals, in particular lambs and piglets, may be due to acute myocarditis. Macroscopical examination of the heart in these cases often reveals a soft, not well-contracted, heart with white-greyish spots or stripes (so-called tiger heart) mainly in the left ventricle and interventricular septum. Often there is an absence of vesicular lesions at the normal predeliction sites in the skin and mucosae, and in hyper-acute cases there may not even be obvious macroscopic lesions in the heart. However, virus can usually be isolated from the myocardium or from the blood and a lympho-histiocytic myocarditis with hyaline degeneration, necrosis of myocytes and infiltration with mononuclear cells can be observed by histopathological examination (Donaldson et al. 1984). Occasionally, the skeletal muscles may also be affected. No lesions can be observed in the myocardium or skeletal muscles of older animals, and virus appears not to replicate in such sites except in young animals (Alexandersen et al., unpublished data). The significance of acute myocarditis in the spread of FMD has not been studied in detail. Death usually occurs rapidly and before the development of vesicular lesions; however, FMDV replication levels are significant in the heart, and it appears plausible that a significant viraemia occurs (Donaldson et al. 1984) and that virus particles are therefore likely to be excreted in breath, saliva, nasal fluid and other excretions and secretions. FMD may also cause abortion in pregnant animals; however, the precipitating fac-

Fig. 4A–D Formalin-fixed, paraffin wax-embedded sections of tissue from pigs inoculated with FMDV O Taiwan 1997 in the heel pad of a left fore foot. **A** Section of tongue 3 days after inoculation. Early vesicular lesions were evident on all four feet and the tongue; however, the area of the tongue shown had no obvious macroscopical lesions. Early microscopical lesions can be seen as microvesicles in the stratum spinosum between and above the papillae. HE, x100. **B** The same section as **A** but at a higher magnification. An early microscopical lesion (microvesicle) can be observed (*arrow*) as a small area of swollen cells with an eosinophilic cytoplasm in the stratum spinosum of the epithelium and a beginning separation and vesicle formation along the basement membrane just sparing the basal cell layer (*arrowhead*). HE, x200. **C** Section of skin from the coronary band of a hind foot at day 3 after inoculation. The tissue was taken at the edge of a macroscopically visible, early vesicular lesion. Swollen cells with eosinophilic cytoplasm are seen (*arrow*) in the stratum spinosum, probably representing early acute cytopathological changes. HE, x400. **D** The same section of skin as that shown in Fig. 3B, but taken at a lower magnification and including the macrocopically visible lesion. Early cellular cytopathology, cell infiltration and vesicle formation are obvious. HE, x200

tors and specific mechanisms involved have not been determined for the various livestock species.

3.7 Mechanisms of Disease

Although the temporal development and appearance of lesions and acute clinical disease are well described, little is actually known about the mechanisms responsible for the severity of the clinical signs. FMD is usually clinically severe in pigs and cattle, but the severity of the disease is not always correlated to the severity or magnitude of vesicular lesions and, in addition, another vesicular virus infection, infection with swine vesicular disease virus (SVDV), causes very similar or identical vesicular lesions in pigs but, in contrast to FMDV, only mild or moderate other clinical signs. Therefore, it appears likely that the FMDV infection induces a general pro-inflammatory reaction leading to additional severe clinical signs, in particular general dullness, inappetence, inability to regulate or maintain body temperature, and in severe cases, death. These aspects of FMD are thought to result from virus-host interactions outside the direct acute cytopathogenic effect of the virus on infected cells and may include cell death (releasing "danger signals"), virus-antibody immune complex formation, complement activation and the production and release of pro-inflammatory and antiviral cytokines, prostaglandins and acute-phase proteins. Cell-free viraemia is usually high in FMD-affected animals and much higher than, for example, in SVDV-infected pigs, and as the initial decrease in viraemia and development of antibodies appears to correlate with the peak of severe clinical disease (Alexandersen et al., unpublished data) we hypothesise that virus-antibody immune complexes may trigger a pro-inflammatory cascade. Studies on the acute-phase protein haptoglobin in FMDV-infected cattle (Hofner et al. 1994) indicated that this acute-phase protein is elevated when viraemia and clinical signs become evident, suggesting that the inflammatory response is indeed activated. We have preliminary data indicating that FMDV infection results in increased concentrations of circulating haptoglobin in pigs and in a complex induction and inhibition of pro-inflammatory and antiviral mRNAs (Murphy, Ahmed, Zhang and Alexandersen, unpublished data).

4
Conclusions

FMD is an important, highly contagious, acute vesicular disease of livestock that can spread very rapidly and affect both domesticated and wild ruminants and pigs. Acute clinical signs may be mild in sheep and goats but are usually associated with severe lameness in pigs and obvious mouth lesions in cattle. FMDV can cause persistent infection in ruminants, so-called carriers, both in ruminants recovered from the acute infection and in vaccinated ruminants if these are subsequently exposed to infectious virus. Much remains to be done to increase our understanding of the determinants specifying host range and mechanisms of pathogenesis.

Acknowledgements We thank Zhidong Zhang, Melvyn Quan, Ciara Murphy, Raza Ahmed and Jeanette Knight for their contributions and Steven Archibald for artwork. The Department for Environment, Food and Rural Affairs (DEFRA), Animal Health Office, Carlisle, is thanked for providing photographs of FMD in cattle. The research was supported by DEFRA and The Biotechnology and Biological Sciences Research Council (BBSRC), UK.

References

Afzal, H. and Barya, M. A. (1968). Occurrence and survival of foot-and-mouth disease virus in external lesions and discharges of experimentally infected buffalo-calves. Bulletin Office International des Epizooties, 69, 509–519

Aggarwal, N., Zhang, Z., Cox, S., Statham, R., Alexandersen, S., Kitching, R. P. and Barnett, P. V. (2002). Experimental studies with foot-and-mouth disease virus, strain O, responsible for the 2001 epidemic in the United Kingdom. Vaccine, 20, 2508–2515

Alexandersen, S., Brotherhood, I. and Donaldson, A. I. (2002a). Natural aerosol transmission of foot-and-mouth disease virus to pigs: minimal infectious dose for strain O1 Lausanne. Epidemiology and Infection, 128, 301–312

Alexandersen, S. and Donaldson, A. I. (2002). Further studies to quantify the dose of natural aerosols of foot-and-mouth disease virus for pigs. Epidemiology and Infection, 128, 313–323

Alexandersen, S., Kitching, R.P., Mansley, L.M. and Donaldson, A.I. (2003a). Clinical and laboratory investigations of five outbreaks during the early stages of the 2001 foot-and-mouth disease epidemic in the United Kingdom. Veterinary Record, 152, 489–496

Alexandersen, S., Oleksiewicz, M. B. and Donaldson, A. I. (2001). The early pathogenesis of foot-and-mouth disease in pigs infected by contact: a quantitative time course study using TaqMan RT-PCR. Journal of General Virology, 82, 747–755

Alexandersen, S., Quan, M., Murphy, C., Knight, J. and Zhang, Z. (2003b). Studies of quantitative parameters of virus excretion and transmission in pigs and cattle experimentally infected with foot-and-mouth disease virus. Journal of Comparative Pathology. In Press

Alexandersen, S., Zhang, Z. and Donaldson, A. (2002b). Aspects of the persistence of foot-and-mouth disease virus in animals—the carrier problem. Microbes and Infection, 4, 1099–1110

Alexandersen, S., Zhang, Z., Donaldson, A.I. and Garland, A.J.M. (2003c). The pathogenesis and diagnosis of foot-and-mouth disease. Journal of Comparative Pathology. In Press

Alexandersen, S., Zhang, Z., Reid, S. M., Hutchings, G. H. and Donaldson, A. I. (2002c). Quantities of infectious virus and viral RNA recovered from sheep and cattle experimentally infected with foot-and-mouth disease virus O UK 2001. Journal of General Virology, 83, 1915–1923

Anderson, E. C., Doughty, W. J., Anderson, J. and Paling, R. (1979). The pathogenesis of foot-and-mouth disease in the African buffalo (Syncerus caffer) and the role of this species in the epidemiology of the disease in Kenya. Journal of Comparative Pathology, 89, 541–549

Armstrong, R., Davie, J., and Hedger, R. S. (1967). Foot-and-mouth disease in man. British Medical Journal, 4, 529–530

Bachrach, H. L. (1968). Foot-and-mouth disease. Annual Review of Microbiology, 22, 201–244

Barnett, P. V. and Cox, S. J. (1999). The role of small ruminants in the epidemiology and transmission of foot-and-mouth disease. Veterinary Journal, 158, 6–13

Bastos, A. D., Boshoff, C. I., Keet, D. F., Bengis, R. G. and Thomson, G. R. (2000). Natural transmission of foot-and-mouth disease virus between African buffalo (Syncerus caffer) and impala (Aepyceros melampus) in the Kruger National Park, South Africa. Epidemiology and Infection., 124, 591–598

Bauer, K. (1997). Foot-and-mouth disease as zoonosis. Archives of Virology, (Suppl), 13, 95–97

Beard, C. W. and Mason, P. W. (2000). Genetic determinants of altered virulence of Taiwanese foot-and-mouth disease virus. Journal of Virology, 74, 987–991

Beck, E. and Strohmaier, K. (1987). Subtyping of European foot-and-mouth disease virus strains by nucleotide sequence determination. Journal of Virology, 61, 1621–1629

Belsham, G. J. (1993). Distinctive features of foot-and-mouth disease virus, a member of the picornavirus family; aspects of virus protein synthesis, protein processing and structure. Progress in Biophysics and Molecular Biology, 60, 241–260

Bengis, R. G., Hedger, R. S., De Vos, V., and Hurter, L. R. (1984). The role of the African elephant Loxodonta africana in the epidemiology of foot-and-mouth disease in the Kruger national park. Proceedings of the 13th World Congress in Buiatrics, 13, 39–44

Black, L., Francis, M. J., Rweyemamu, M. M., Umebara, O. and Boge, A. (1984). The relationship between serum antibody titres and protection from foot and mouth disease in pigs after oil emulsion vaccination. Journal of Biological Standardization, 12, 379–389

Brown, F. and Cartwright, B. (1960). Purification of the virus of foot-and-mouth disease by fluorocarbon treatment and its dissociation from neutralizing antibody. Journal of Immunology, 85, 309–313
Burrows, R. (1966a). Studies on the carrier state of cattle exposed to foot-and-mouth disease virus. Journal of Hygiene (London), 64, 81–90
Burrows, R. (1966b). The infectivity assay of foot-and-mouth disease virus in pigs. Journal of Hygiene (London), 64, 419–429
Burrows, R. (1968a). Excretion of foot-and-mouth disease virus prior to the development of lesions. Veterinary Record, 83, 387–388
Burrows, R. (1968b). The persistence of foot-and mouth disease virus in sheep. Journal of Hygiene (London), 66, 633–640
Burrows, R., Mann, J. A., Garland, A. J., Greig, A. and Goodridge, D. (1981). The pathogenesis of natural and simulated natural foot-and-mouth disease infection in cattle. Journal of Comparative Pathology, 91, 599–609
Callens, M., De Clercq, K., Gruia, M. and Danes, M. (1998). Detection of foot-and-mouth disease by reverse transcription polymerase chain reaction and virus isolation in contact sheep without clinical signs of foot-and-mouth disease. Veterinary Quarterly, 20 Suppl 2, 37–40
Capel-Edwards, M. (1967). Foot-and-mouth disease in Myocastor coypus. Journal of Comparative Pathology, 77, 217–221
Capel-Edwards, M. (1969). Spread of foot-and-mouth disease. Lancet, 2, 901
Capel-Edwards, M. (1970). Foot-and-mouth disease in the brown rat. Journal of Comparative Pathology, 80, 543–548
Capel-Edwards, M. (1971a). The susceptibility of small mammals to foot and mouth disease virus. The Veterinary Bulletin, 41, 815–823
Capel-Edwards, M. (1971b). The susceptibility of three British small mammals to foot-and-mouth disease. Journal of Comparative Pathology, 81, 433–436
Condy, J. B., Hedger, R. S., Hamblin, C. and Barnett, I. T. (1985). The duration of the foot-and-mouth disease virus carrier state in African buffalo (i) in the individual animal and (ii) in a free-living herd. Comparative Immunology and Microbiology of Infectious Diseases, 8, 259–265
Cottral, G. E. and Bachrach, H. L. (1968). Foot-and-mouth disease viremia. Proceedings of the Annual Meeting of the United States Animal Health Association, 72, 383–399
Cottral, G. E., Gailiunas, P. and Campion, R. L. (1963). Detection of foot-and-mouth disease virus in lymph nodes of cattle throughout course of infection. Proceedings of the Annual Meeting of the United States Livestock Sanitary Association, 67, 463–472
Cottral, G. E., Patty, R. E., Gailiunas, P. and Scott, F. W. (1965). Sensitivity of cell cultures, cattle, mice, and guinea-pigs for detection of nineteen foot-and-mouth disease viruses. Bulletin Office International Epizooties, 63, 1607–1625
Donaldson, A. I. (1986). Aerobiology of foot-and-mouth disease (FMD): an outline and recent advances. Revue Scientifique et Technique l'Office International des Épizooties, 5, 315–321
Donaldson, A. I. (1998). Experimental and natural adaptation of strains of foot-and-mouth disease virus to different species. Session of the Research Group of the

Standing Technical Committee, European Commission for the Control of Foot-and-Mouth Disease, pp, 18–22
Donaldson, A. I. and Alexandersen, S. (2001). The relative resistance of pigs to infection by natural aerosols of foot-and-mouth disease virus. Veterinary Record, 148, 600–602
Donaldson, A. I. and Alexandersen, S. (2002). Predicting the spread of foot and mouth disease by airborne virus. Revue Scientifique et Technique l'Office International des Épizooties, 21, 569–575
Donaldson, A. I., Alexandersen, S., Sorensen, J. H. and Mikkelsen, T. (2001). The relative risks of the uncontrollable (airborne) spread of foot-and-mouth disease by different species. Veterinary Record, 148, 602–604
Donaldson, A. I. and Ferris, N. P. (1980). Sites of release of airborne foot-and-mouth disease virus from infected pigs. Research in Veterinary Science, 29, 315–319
Donaldson, A. I., Ferris, N. P. and Gloster, J. (1982a). Air sampling of pigs infected with foot-and-mouth disease virus: comparison of Litton and cyclone samplers. Research in Veterinary Science, 33, 384–385
Donaldson, A. I., Ferris, N. P. and Wells, G. A. (1984). Experimental foot-and-mouth disease in fattening pigs, sows and piglets in relation to outbreaks in the field. Veterinary Record, 115, 509–512
Donaldson, A. I., Gibson, C. F., Oliver, R., Hamblin, C. and Kitching, R. P. (1987). Infection of cattle by airborne foot-and-mouth disease virus: minimal doses with O1 and SAT 2 strains. Research in Veterinary Science, 43, 339–346
Donaldson, A. I., Gloster, J., Harvey, L. D. and Deans, D. H. (1982b). Use of prediction models to forecast and analyse airborne spread during the foot-and-mouth disease outbreaks in Brittany, Jersey and the Isle of Wight in 1981. Veterinary Record, 110, 53–57
Donaldson, A. I., Herniman, K. A., Parker, J. and Sellers, R. F. (1970). Further investigations on the airborne excretion of foot-and-mouth disease virus. Journal of Hygiene (London), 68, 557–564
Donaldson, A. and Knowles, N. (2001). Foot-and-mouth disease in man. Veterinary Record, 148, 319
Donaldson, A. I. and Sellers, R. F. (2000). Foot-and-mouth disease. In: Diseases of Sheep, 3rd Edit., W. B. Martin and I. D. Aitken, Eds., Blackwell Science, Oxford, pp. 254–258
Dunn, C. S. and Donaldson, A. I. (1997). Natural adaption to pigs of a Taiwanese isolate of foot-and-mouth disease virus. Veterinary Record, 141, 174–175
Dutta, P. K., Sarma, G., and Das, S. K. (1983). Foot-and-mouth disease in Indian buffaloes. Veterinary Record, 113, 134
Farag, M. A., Al-Sukayran, A., Mazloum, K. S., and Al-Bokmy, A. M. (1998). The susceptibility of camels to natural infection with foot and mouth disease virus. Assiut Veterinary Medical Journal, 40, 201–211
Fondevila, N. A., Marcoveccio, F. J., Blanco Viera, J. B., O'Donnell, V. K., Carrillo, B. J., Schudel, A. A., David, M., Torres, A., and Mebus, C. A. (1995). Susceptibility of llamas (Lama glama) to infection with foot-and-mouth-disease virus. Zentralblatt fur Veterinarmedicine B, 42, 595–599
Forman, A. J. and Gibbs, E. P. (1974). Studies with foot-and-mouth disease virus in British deer (red, fallow and roe). Journal of Comparative Pathology, 84, 215–220

Forman, A. J., Gibbs, E. P., Baber, D. J., Herniman, K. A., and Barnett, I. T. (1974). Studies with foot-and-mouth disease virus in British deer (red, fallow and roe). II. Recovery of virus and serological response. Journal of Comparative Pathology, 84, 221–229

Francis, M. J. and Black, L. (1983). Antibody response in pig nasal fluid and serum following foot-and-mouth disease infection or vaccination. Journal of Hygiene (London), 91, 329–334

Francis, M. J., Ouldridge, E. J. and Black, L. (1983). Antibody response in bovine pharyngeal fluid following foot-and-mouth disease vaccination and, or, exposure to live virus. Research in Veterinary Science, 35, 206–210

Gailiunas, P. (1968). Microscopic skin lesions in cattle with foot-and-mouth disease. Archives Gesamte Virusforschung, 25, 188–200

Garland, A. J. (1974). The inhibitory activity of secretions in cattle against FMDV. PhD Thesis, University of London

Garland, A. J. M. and Donaldson, A. I. (1990). Foot-and-mouth disease. Surveillance, 17, 6–8

Gibbs, E. P., Herniman, K. A., Lawman, M. J. and Sellers, R. F. (1975a). Foot-and-mouth disease in British deer: transmission of virus to cattle, sheep and deer. Veterinary Record, 96, 558–563

Gibbs, E. P., Herniman, K. A. and Lawman, M. J. (1975b). Studies with foot-and-mouth disease virus in British deer (muntjac and sika). Clinical disease, recovery of virus and serological response. Journal ofComparative Pathology, 85, 361–366

Gibson, C. F. and Donaldson, A. I. (1986). Exposure of sheep to natural aerosols of foot-and-mouth disease virus. Research in Veterinary Science, 41, 45–49

Gloster, J., Blackall, J., Sellers, R. F. and Donaldson, A. I. (1981). Forecasting the spread of foot-and-mouth disease. Veterinary record, 108, 370–374

Gloster, J., Sellers, R. F. and Donaldson, A. I. (1982). Long distance transport of foot-and-mouth disease virus over the sea. Veterinary Record, 110, 47–52

Hamblin, C., Kitching, R. P., Donaldson, A. I., Crowther, J. R. and Barnett, I. T. (1987). Enzyme-linked immunosorbent assay (ELISA) for the detection of antibodies against foot-and-mouth disease virus. III. Evaluation of antibodies after infection and vaccination. Epidemiology and Infection, 99, 733–744

Hatch, T. F. and Gross, P. (1964) Pulmonary Deposition and Retention of Inhaled Aerosols, Academic Press, London & New York, pp. 45–68

Hedger, R. S. (1972). Foot-and-mouth disease and the African buffalo (Syncerus caffer). Journal of Comparative Pathology, 82, 19–28

Hedger, R. S. and Brooksby, J. B. (1976). FMD in an Indian elephant. Veterinary Record, 99, 93

Hedger, R. S. and Condy, J. B. (1985). Transmission of foot-and-mouth disease from African buffalo virus carriers to bovines. Veterinary record, 117, 205

Hedger, R. S., Condy, J. B. and Golding, S. M. (1972). Infection of some species of African wild life with foot-and-mouth disease virus. Journal of Comparative Pathology, 82, 455–461

Henderson, W. M. (1948). Further consideration of some of the factors concerned in intracutaneous injection of cattle. Journal of Pathology and Bacteriology, 60, 137–139

Henderson, W. M. (1949). The Quantitative Study of Foot-and-Mouth Disease Virus, Her Majesty's Stationary Office, London, p. 8
Henderson, W. M. (1952). A comparison of different routes of inoculation of cattle for detection of the virus of foot-and-mouth disease. Journal of Hygiene (London), 50, 182–194
Hess, W. R., Bachrach, H. L. and Callis, J. J. (1967). Persistence of foot-and-mouth disease virus in bovine kidneys and blood as related to the occurrence of antibodies. American Journal of Veterinary Research, 21, 1104–1108
Hofner, M. C., Fosbery, M. W., Eckersall, P. D. and Donaldson, A. I. (1994). Haptoglobin response of cattle infected with foot-and-mouth disease virus. Research in Veterinary Science, 57, 125–128
Howell, P. G., Young, E. and Hedger, R. S. (1973). Foot-and-mouth disease in the African elephant (Loxodonta africana). Onderstepoort Journal of Veterinary Research, 40, 41–52
Hughes, G. J., Mioulet, V., Kitching, R. P., Woolhouse, M. E., Alexandersen, S. and Donaldson, A. I. (2002). Foot-and-mouth disease virus infection of sheep: implications for diagnosis and control. Veterinary Record, 150, 724–727
Hyslop, N. S. (1970). The epizootiology and epidemiology of foot and mouth disease. Advances in Veterinary Science and Comparative Medicine, 14, 261–307
Jackson, T., Blakemore, W., Newman, J. W., Knowles, N. J., Mould, A. P., Humphries, M. J. and King, A. M. (2000a). Foot-and-mouth disease virus is a ligand for the high-affinity binding conformation of integrin $\alpha 5\beta 1$: influence of the leucine residue within the RGDL motif on selectivity of integrin binding. Journal of General Virology, 81 Pt 5, 1383–1391
Jackson, T., Mould, A. P., Sheppard, D. and King, A. M. (2002). Integrin $\alpha v\beta 1$ is a receptor for foot-and-mouth disease virus. Journal of Virology, 76, 935–941
Jackson, T., Sharma, A., Ghazaleh, R. A., Blakemore, W. E., Ellard, F. M., Simmons, D. L., Newman, J. W., Stuart, D. I. and King, A. M. (1997). Arginine-glycine-aspartic acid-specific binding by foot-and-mouth disease viruses to the purified integrin $\alpha(v)\beta 3$ in vitro. Journal of Virology, 71, 8357–8361
Jackson, T., Sheppard, D., Denyer, M., Blakemore, W. and King, A. M. (2000b). The epithelial integrin $\alpha v\beta 6$ is a receptor for foot-and-mouth disease virus. Journal of Virology, 74, 4949–4956
Jerez, J. A., Pinto, A. A., Arruda, N. V., Koseki, I., Abuhab, T. G., and Rodrigues, M. A. (1979). [Foot-and-mouth disease in buffaloes (Bubalus bubalis, Linnaeus, 1758): search of antiantigen antibodies and isolation of the virus]. Arquivos do Instituto Biologico (Sao Paulo), 46, 111–115
King, A. M. Q., Brown, F., Christian, P., Hovi, T., Hyypia, T., Knowles, N. J., Lemon, S. M., Minor, P. D., Palmenberg, A. C., Skern, T., and Stanway, G. (2000). "Picornaviridae," in Virus taxonomy: classification and nomenclature of viruses: seventh report of the International Committee on Taxonomy of Viruses, M. H. V. van Regenmortel, C. M. Fauquet, & D. H. L. Bishop, eds., Academic Press, London, pp. 657–678
Kitching, R. P. (1998). A recent history of foot-and-mouth disease. Journal of Comparative Pathology, 118, 89–108

Kitching, R. P. and Alexandersen, S. (2002). Clinical variation in foot and mouth disease: pigs. Revue Scientifique et Technique l'Office International des Épizooties, 21, 513–518

Kitching, R. P., Knowles, N. J., Samuel, A. R. and Donaldson, A. I. (1989). Development of foot-and-mouth disease virus strain characterisation—a review. Tropical Animal Health and Production, 21, 153–166

Knowles, N. J., Davies, P. R., Henry, T., O'Donnell, V., Pacheco, J. M. and Mason, P. W. (2001a). Emergence in Asia of foot-and-mouth disease viruses with altered host range: characterization of alterations in the 3A protein. Journal of Virology, 75, 1551–1556

Knowles, N. J., Samuel, A. R., Davies, P. R., Kitching, R. P. and Donaldson, A. I. (2001b). Outbreak of foot-and-mouth disease virus serotype O in the UK caused by a pandemic strain. Veterinary record, 148, 258–259

McCullough, K. C., De Simone, F., Brocchi, E., Capucci, L., Crowther, J. R. and Kihm, U. (1992). Protective immune response against foot-and-mouth disease. Journal of Virology, 66, 1835–1840

McKenna, T. S., Lubroth, J., Rieder, E., Baxt, B. and Mason, P. W. (1995). Receptor binding site-deleted foot-and-mouth disease (FMD) virus protects cattle from FMD. Journal of Virology, 69, 5787–5790

McLauchlan, J. D. and Henderson, W. M. (1947). The occurrence of foot-and-mouth disease in the hedgehog under natural conditions. Journal of Hygiene (Lond), 45, 474–479

McVicar, J. W. and Sutmoller, P. (1969a). Sheep and goats as foot-and-mouth disease carriers. Proceedings of the Annual Meeting of the United States Livestock Sanitary Association., pp. 400–406

McVicar, J. W. and Sutmoller, P. (1969b). The epizootiological importance of foot-and-mouth disease carriers. II. The carrier status of cattle exposed to foot-and-mouth disease following vaccination with an oil adjuvant inactivated virus vaccine. Archives Gesamte Virusforschung, 26, 217–224

McVicar, J. W. and Sutmoller, P. (1974). Neutralizing activity in the serum and oesophageal-pharyngeal fluid of cattle after exposure to foot-and-mouth disease virus and subsequent re-exposure. Archives Gesamte Virusforschung, 44, 173–176

McVicar, J. W. and Sutmoller, P. (1976). Growth of foot-and-mouth disease virus in the upper respiratory tract of non-immunized, vaccinated, and recovered cattle after intranasal inoculation. Journal of Hygiene (London), 76, 467–481

McVicar, J. W., Sutmoller, P., Ferris, D. H. and Campbell, C. H. (1974). Foot-and-mouth disease in white-tailed deer: clinical signs and transmission in the laboratory. Proceedings of the Annual Meeting of the United States Animal Health Association, pp. 169–180

Moussa, A. A., Daoud, A., Tawfik, S., Omar, A., Azab, A. and Hassan, N. A. (1979). Susceptibility of water-buffaloes to infection with foot-and-mouth disease virus. Journal of the Egyptian Veterinary Medical Association, 39, 65–83

Moussa, A. A., Daoud, A., Omar, A., Metwally, N., El-Nimr, M., and McVicar, J. W. (1987). Isolation of FMDV from camels with ulcerative disease syndromes. Journal of the Egyptian Veterinary Medical Association, 47, 219–229

Murphy, M. L., Forsyth, M. A., Belsham, G. J. and Salt, J. S. (1999). Localization of foot-and-mouth disease virus RNA by in situ hybridization within bovine tissues. Virus Research, 62, 67–76

Neff, S., Mason, P. W. and Baxt, B. (2000). High-efficiency utilization of the bovine integrin $\alpha_v\beta_3$ as a receptor for foot-and-mouth disease virus is dependent on the bovine β_3 subunit. Journal of Virology, 74, 7298–7306

Neff, S., Sa-Carvalho, D., Rieder, E., Mason, P. W., Blystone, S. D., Brown, E. J. and Baxt, B. (1998). Foot-and-mouth disease virus virulent for cattle utilizes the integrin $\alpha_v\beta_3$ as its receptor. Journal of Virology, 72, 3587–3594

Newman, J. F., Rowlands, D. J., and Brown, F. (1973). A physico-chemical sub-grouping of the mammalian picornaviruses. Journal of General Virology, 18, 171–180

Oleksiewicz, M. B., Donaldson, A. I. and Alexandersen, S. (2001). Development of a novel real-time RT-PCR assay for quantitation of foot-and-mouth disease virus in diverse porcine tissues. Journal of Virological Methods, 92, 23–35

Oliver, R. E., Donaldson, A. I., Gibson, C. F., Roeder, P. L., Blanc Smith, P. M. and Hamblin, C. (1988). Detection of foot-and-mouth disease antigen in bovine epithelial samples: comparison of sites of sample collection by an enzyme linked immunosorbent assay (ELISA) and complement fixation test. Research in Veterinary Science, 44, 315–319

Piragino, S. (1970). FMD in a circus elephant. Zooprofilassi, 25, 17–22

Platt, H. (1956). A study of the pathological changes produced in young mice by the virus of foot-and-mouth disease. Journal of Pathology and Bacteriology, 122, 299–312

Puntel, M., Fondevila, N. A., Blanco, V. J., O'Donnell, V. K., Marcovecchio, J. F., Carrillo, B. J., and Schudel, A. A. (1999). Serological survey of viral antibodies in llamas (Lama glama) in Argentina. Zentralblatt fur Veterinarmedicine B, 46, 157–161

Rieder, E., Baxt, B. and Mason, P. W. (1994). Animal-derived antigenic variants of foot-and-mouth disease virus type A12 have low affinity for cells in culture. Journal of Virology, 68, 5296–5299

Rieder, E., Berinstein, A., Baxt, B., Kang, A. and Mason, P. W. (1996). Propagation of an attenuated virus by design: engineering a novel receptor for a noninfectious foot-and-mouth disease virus. Proceedings of the National Academy of Science USA, 93, 10428–10433

Sa-Carvalho, D., Rieder, E., Baxt, B., Rodarte, R., Tanuri, A. and Mason, P. W. (1997). Tissue culture adaptation of foot-and-mouth disease virus selects viruses that bind to heparin and are attenuated in cattle. Journal of Virology, 71, 5115–5123

Salt, J. S. (1998). Persistent infection with foot-and-mouth disease virus. Topics in Tropical Virology, 1, 77–128

Sahu, S. P. and Dardiri, A. H. (1979). Susceptibility of mink to certain viral animal diseases foreign to the United States. Journal of Wildlife Diseases, 15, 489–494

Sellers, R. F. (1971). Quantitative aspects of the spread of foot and mouth disease. Veterinary Bulletin, 41, 431–439

Sellers, R. F., Donaldson, A. I., and Herniman, K. A. (1970). Inhalation, persistence and dispersal of foot-and-mouth disease virus by man. Journal of Hygiene (Lond), 68, 565–573

Sellers, R. F. and Forman, A. J. (1973). The Hampshire epidemic of foot-and-mouth disease, 1967. Journal of Hygiene (London), 71, 15–34

Sellers, R. F. and Parker, J. (1969). Airborne excretion of foot-and-mouth disease virus. Journal of Hygiene (London), 67, 671–677

Shimshony, A., Orgad, U., Baharav, D., Prudovsky, S., Yakobson, B., Bar, M. B. and Dagan, D. (1986). Malignant foot-and-mouth disease in mountain gazelles. Veterinary Record, 119, 175–176

Skinner, H. H. (1951). Propagation of strains of foot-and-mouth disease in unweaned white mice. Proceedings of the Royal Society of Medicine, 44, 1041–1044

Skinner, H. H. (1952). Use of unweaned white mice in foot-and-mouth disease research. Nature, 169, 794–797

Skinner, H. H. (1953). One week old white mice as test animals in foot-and-mouth disease research. Proceedings from XVth International Veterinary Congress, IB45, 3–8

Skinner, H. H. (1954). Infection of chickens and chick embryos with the viruses of foot-and-mouth disease and of vesicular stomatitis. Nature, 174, 1052–1054

Snowdon, W. A. (1968). The susceptibility of some Australian fauna to infection with foot and mouth disease virus. Australian Journal of Experimental Biology and Medical Science, 46, 667–687

Sorensen, J. H., Jensen, C. O., Mikkelsen, T., Mackay, D. K. and Donaldson, A. I. (2001). Modelling the atmospheric dispersion of foot-and-mouth disease virus for emergency preparedness. Physics Chemistry Earth, 26, 93–97

Sorensen, J. H., Mackay, D. K., Jensen, C. O. and Donaldson, A. I. (2000). An integrated model to predict the atmospheric spread of foot-and-mouth disease virus. Epidemiology and Infection, 124, 577–590

Sutmoller, P. and Gaggero, A. (1965). Foot-and mouth disease carriers. Veterinary Record, 77, 968–969

Sutmoller, P., McVicar, J. W. and Cottral, G. E. (1968). The epizootiological importance of foot-and-mouth disease carriers. I. Experimentally produced foot-and-mouth disease carriers in susceptible and immune cattle. Archives Gesamte Virusforschung, 23, 227–235

Thomson, G. R. (1994). Foot-and-mouth disease. In: Infectious Diseases of Livestock with Special Reference to Southern Africa, J. A. W. Coetzer G. R. Thomsen, R. C. Tustin, and N. P. J. Kriek, eds., Oxford University Press, Cape Town, pp. 825–852

Thomson, G. R., Bengis, R. G., Esterhuysen, J. J. and Pini, A. (1984). Maintenance mechanisms for foot-and-mouth disease virus in the Kruger national park and potential avenues for its escape into domestic animal populations. Proceedings of the 13th World Congress in Buiatrics, 13, 33–38

Thomson, G. R., Vosloo, W. and Bastos, A. D. (2003). Foot and mouth disease in wildlife. Virus Research, 91,145–161

Turbitt, D. (2001). No human cases so far in foot and mouth epidemic in the United Kingdom. Eurosurveillance Weekly, 5, 1

Van Bekkum, J. G., Frenkel, H. S., Frederiks, H. H. J. and Frenkel, S. (1959a). Observations on the carrier state of cattle exposed to foot-and-mouth disease virus. Bulletin Office International Epizooties, 51, 917–922

Van Bekkum, J. G., Frenkel, H. S., Frederiks, H. H. J. and Frenkel, S. (1959b). Observations on the carrier state of cattle exposed to foot-and-mouth disease virus. Tijdschrieft fur Diergeneeskunde, 84, 1159–1164

Yilma, T. (1980). Morphogenesis of vesiculation in foot-and-mouth disease. American Journal of Veterinary Research, 41, 1537–1542

Zhang, Z. and Kitching, P. (2000). A sensitive method for the detection of foot and mouth disease virus by in situ hybridisation using biotin-labelled oligodeoxynucleotides and tyramide signal amplification. Journal of Virological Methods, 88, 187–192

Zhang, Z. D. and Kitching, R. P. (2001). The localization of persistent foot and mouth disease virus in the epithelial cells of the soft palate and pharynx. Journal of Comparative Pathology, 124, 89–94

CTMI (2005) 288:43–70

Translation and Replication of FMDV RNA

G. J. Belsham

BBSRC Institute for Animal Health, Pirbright, Woking, Surrey, GU24 ONF, UK
graham.belsham@bbsrc.ac.uk

Abstract Foot-and-mouth disease virus (FMDV) RNA is infectious. After delivery of the RNA (about 8.3 kb) into the cytoplasm of a cell, the RNA must initially be translated to produce the viral proteins required for RNA replication and for the packaging of the RNA into new virions. Subsequently there has to be a switch in the function of the RNA; translation has to be stopped to permit RNA replication. The signals required for the control of the different roles of viral RNA must be included within the viral RNA sequence. Many cellular proteins interact with the viral RNA and probably also with the virus-encoded proteins. The functions of different RNA elements within the viral RNA and the various virus-encoded proteins in determining the efficiency of virus replication are discussed. Unique aspects of FMDV RNA translation and replication are emphasised.

1 Introduction

Foot-and-mouth disease virus (FMDV) is a member of the picornavirus family. The picornaviruses are currently divided into nine genera; FMDV is the prototype aphthovirus. Other well-known picornaviruses include poliovirus (PV, an enterovirus), human rhinoviruses and encephalomyocarditis virus (EMCV, a cardiovirus). Picornavirus particles are roughly spherical (about 30 nm in diameter) and are comprised of 60 copies of 4 different virus-encoded capsid proteins, VP1 (1D), VP2 (1B), VP3 (1C) and VP4 (1A) together with a single copy of the viral RNA genome (see Fig. 1). Each picornavirus has a single-stranded RNA genome of positive polarity (about 8 kb in length) and the genomic RNA is infectious (see, e.g. Belsham and Bostock 1988). The virus particle serves to deliver just the virus genome into the cytoplasm of a cell. It follows that the first step in the replication of the virus is the translation of the viral RNA to produce each of the virus-encoded proteins. These proteins are required for viral RNA replication and for the production of new virus particles which are able to initiate a fresh cycle of infection. Some of the viral proteins also modify the functions of the host cell in which the virus is replicating. These changes serve to block host-defense mechanisms and/or to facilitate the replication of the virus.

Because viruses use the host-cell machinery, it is apparent that there must be many interactions between host cell proteins and the viral proteins and RNA. Some of these interactions are well characterised, but many others remain to be identified. These interactions undoubtedly play an important role in determining which cell types are efficiently infected by the virus and hence the outcome of the infection in terms of disease. This chapter describes features of the viral RNA and the proteins it encodes which influence the replication of the virus.

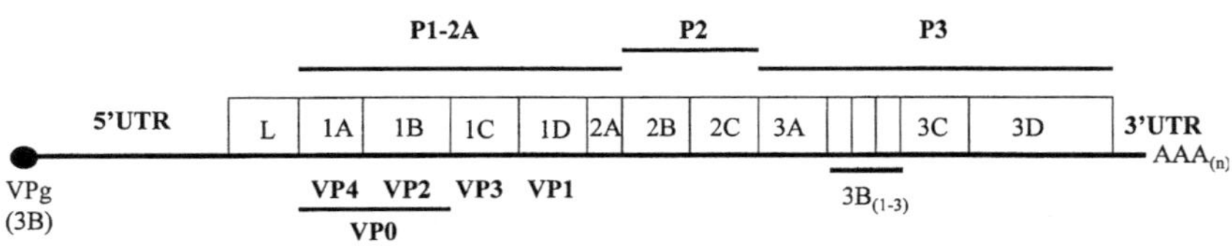

Fig. 1 Genome organisation of FMDV. The non-coding and coding regions of FMDV RNA are indicated. The single large open reading frame encodes a polyprotein which is never observed. It is processed by virus-encoded proteases during synthesis. Many different precursors can be generated during this processing of the polyprotein; some of the major precursors are indicated

2
Structure and Function of FMDV RNA

2.1
Virion RNA

FMDV genomic RNA is about 8.3 kb in length (see Fig. 1). All of the viral RNA is of the same length; no sub-genomic mRNAs are produced by picornaviruses. The viral genome has some similarities to a eukaryotic cell mRNA in that it contains a single long open reading frame (ORF; the FMDV ORF is about 7000 nt long) followed by a 3′ -untranslated region (UTR; about 100 nt) and a poly(A) tail. However, it also has certain characteristics which are very different from cellular mRNAs. Typically, cellular mRNAs have a 5′-UTR of about 50–100 nt. A feature of the FMDV RNA is the presence of a very long 5′-UTR of about 1300 nt. The RNAs from other picornaviruses also have long 5′-UTRs, but the FMDV 5′-UTR is much larger than most (e.g. the PV 5′-UTR is about 740 nt and the EMCV 5′-UTR is about 830 nt). All cytoplasmic eukaryotic mRNAs have a cap structure [m^7GpppN..., where m is a methyl group and the bond between the methylated G residue to the next base (N) occurs by a 5'-5' linkage] at their 5′ terminus. This structure is required for the efficient recognition of the mRNA by the translation initiation complex eIF4F (see below and Gingras et al. 1999 for review). The viral RNA lacks this cap structure; indeed, a short virus-encoded peptide, termed VPg (or 3B), is covalently attached to the 5′ terminus of the genomic RNA. However, this modification is rapidly lost within cells and hence much of the viral RNA within infected cells has a free 5′ end.

3
Features of the 5′-UTR

3.1
The S-Fragment

The 5′-UTR of FMDV RNA contains several discrete regions (Fig. 2). The first portion (ca. 350 nt) is called the S-fragment and is predicted to fold into a large hairpin structure (Clarke et al. 1987; Escarmis et al. 1992), but its function is unknown. It is assumed to be required for RNA replication, but it has been little studied. In contrast, at the 5′-end of the PV RNA is a much better-characterised 'cloverleaf' structure. This structure is only about 80 nt long and has been shown to interact with both cellu-

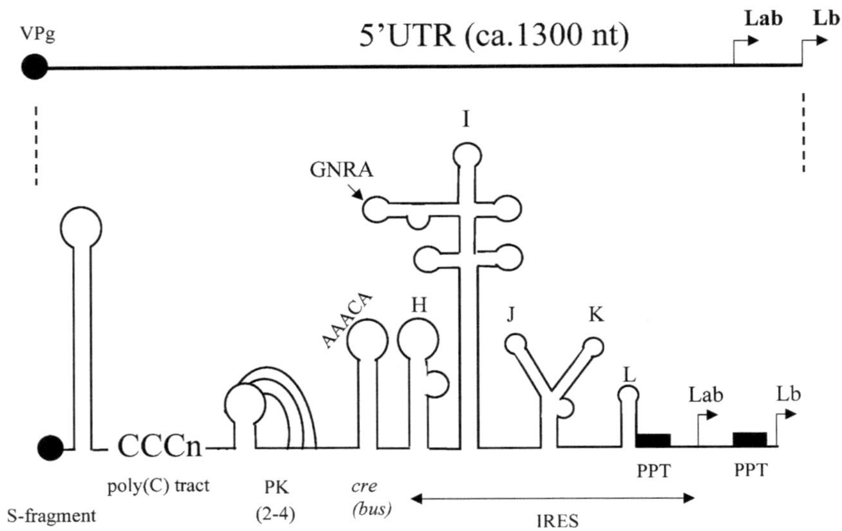

Fig. 2 Representation of the structural elements within the FMDV 5′-UTR. The 5′-UTR is predicted to contain extensive secondary structure. Some of the elements have defined functions, but others do not. Note that the stems indicated are not perfectly base-paired throughout. See the text for information on studies on the detailed analysis of these RNA structures. Individual domains within the IRES are labelled *(H–L)* and are referred to within the text. The conserved GNRA motif and the polypyrimidine tracts (*PPTs*) are indicated. Abbreviations: *PK*, pseudoknot; *cre* (*cis*-acting replication element)/bus (3B-uridylylation site); *IRES*, internal ribosome entry site. The two different start sites (84 nt apart) for the polyprotein that give rise to Lab and Lb are shown

lar and viral proteins (Garmarnik and Andino 1997; Parsley et al. 1997). It is involved in the process of RNA replication (Andino et al. 1990) and also has a major effect on the stability of the viral RNA (Murray et al. 2001).

3.2
Poly(C) Tract and Pseudoknots

The presence of a poly(C) tract within the 5′-UTR (see Fig. 2) is common to the aphthoviruses [FMDV and equine rhinitis A virus (ERAV)] and most cardioviruses (except Theiler's murine encephalitis virus). Typically field strains of FMDV have a poly(C) tract of about 150–200 nt (Brown et al. 1974; Harris and Brown 1977). Within laboratory strains, the tract can be rather shorter (ca. 80 nt) but the longest poly(C) tract identified was found in a strain of FMDV recovered from persistently infected BHK cells; in this case a tract of about 450 nt was identified (Escarmis et al. 1992). The importance of this region is demonstrated by the dramatic increase in the length of the poly(C) tract that occurs when FMDV is rescued from RNA transcripts derived from cloned viral cDNA. When the plasmid contained just 6 C residues at this site, it was found that the rescued viruses contained poly(C) tracts with at least 80 C residues (Rieder et al. 1993). However, when just 2 C residues were present within the plasmid, the tract was not amplified in the rescued virus. This mutant virus grew rather slowly in tissue culture but remained as pathogenic in mice as viruses with longer poly(C) tracts (Rieder et al. 1993). It should be noted that within cardiovirus RNAs the length of the poly(C) tract seems to be stable. It has been shown that mengoviruses containing a genome with a short tract are attenuated in mice (Duke et al. 1990); however, this was not true for the closely related EMCV (Hahn and Palmenberg 1995); thus the role of this feature within the genomes of different picornaviruses is rather unclear.

On the 3′ side of the poly(C) tract within FMDV RNA is a region which is predicted to contain multiple (2–4) pseudoknots (Clarke et al. 1987; Escarmis et al. 1995). The role of these elements is not known, but it may be significant that in cardiovirus RNAs such elements are located on the 5′ side of the poly(C) tract (Martin and Palmenberg 1996); hence it is possible that the pseudoknots have some role in association with this tract.

3.3 '*cis*-Acting Replication Element'

Recently, Mason et al. (2002) have demonstrated that the FMDV 5′-UTR contains a stable stem-loop element of about 55 nt which has properties similar to structures found in other picornavirus RNAs. These elements were first discovered when it was shown that a region within the P1-coding sequence of the HRV-14 genome was required to permit replication of an RNA transcript (McKnight and Lemon 1996, 1998). The element necessary to permit replication was required as RNA and was predicted to form a stable stem-loop structure which was termed a '*cis*-acting replication element' (*cre*). Analogous elements have now been identified in several different picornavirus RNAs. Each of the picornavirus *cre* structures contains a conserved motif of AAACA located within the loop region. This motif acts as the template for the uridylylation of VPg (3B) by the viral RNA polymerase to produce VPgpU and/or VPgpUpU (Paul et al. 2000; Gerber et al. 2001). These products act as the primers for the initiation of viral RNA synthesis, thus explaining the presence of VPg at the 5′ terminus of both positive- and negative-sense RNA transcripts. Amongst the various picornavirus RNAs studied, the *cre* structures have been identified in different locations within the genome. For example, the HRV-14 *cre* is located within the coding sequence for VP1 (McKnight and Lemon 1998) whereas the PV *cre* is located within the 2C sequence (Goodfellow et al. 2000). Indeed, FMDV is apparently unique in having this structure located outside of the coding region and within the 5′-UTR (see Fig. 2). However, it has been demonstrated that these elements can be moved within the genome and still retain activity at their new locations (see, e.g. Goodfellow et al. 2000; Mason et al. 2002).

The identification of the FMDV *cre* within the 5′-UTR has provided an explanation for some observations on a temperature-sensitive (*ts*) mutant of FMDV. The replication of this virus is greatly suppressed at the non-permissive temperature and, surprisingly, sequence analysis identified the location of the *ts* mutation as within the stem-loop structure within the 5′-UTR now identified as the *cre* (see Tiley et al. 2003). Importantly, it was possible to complement the defect in its replication at the non-permissive temperature by co-infection with other *ts* mutant FMDVs that were also defective in RNA replication under these conditions but had defects in other regions of the genome. These data indicate that the FMDV '*cis*-acting replication element' can function *in trans*. This result is consistent with the observation that a pool of free VPgpUpU (the product of the *cre*-templated uridylylation of VPg) is generated within PV-infected cells (Crawford and Baltimore 1983), and hence it seems reasonable that these elements should be able to function *in trans*. Thus the term *cre* seems inappropriate. Because a specific function for the element has now been defined (Paul et al. 2000; Gerber et al. 2001), it has been suggested that the term *cre* should be replaced by '3B-uridylylation site' (*bus)* (see Tiley et al. 2003).

3.4 The Internal Ribosome Entry Site (IRES)

The 3′ portion of the FMDV 5′-UTR is required for the initiation of protein synthesis on the viral RNA. The features of the FMDV 5′-UTR made it very unlikely that FMDV RNA could be translated by the classic cap-dependent translation initiation mechanism. However, FMDV RNA is a very efficient template for translation. The absence of a cap structure and the presence of an extensive secondary structure with multiple unused AUG codons are features shared by all picornavirus RNA 5′-UTRs. A major step forward in understanding the mechanism of picornavirus translation initiation was achieved when Pelletier and Sonenberg (1988) demonstrated that the PV 5′-UTR was able to direct cap-independent internal initiation of protein synthesis. Analogous results were also obtained with the 5′-UTR from EMCV by Jang et al. (1988). The element required for this activity is now usually referred to as an internal ribosome entry site (IRES). Shortly afterwards, it was demonstrated that an element (located immediately upstream of the polyprotein coding region) of about 450 nt within the 5′-UTR of FMDV RNA functioned as an IRES (Belsham and Brangwyn 1990; Kuhn et al. 1990).

The classic assay for IRES elements is the construction and analysis of dicistronic mRNAs in which the putative IRES element is introduced between two reporter sequences (see Belsham and Sonenberg 1996, 2000; Belsham and Jackson 2000). IRES elements have been defined for seven of the nine genera of picornaviruses to date. The activity of most picornavirus IRES elements is, at least, maintained when cap-dependent protein synthesis is blocked either by the cleavage of the translation initiation factor eIF4G (see below) or by the sequestration of eIF4E (the cap-binding protein) with 4E-BP1 (Pause et al. 1994a; Roberts et al. 1998). However, this is not true for the hepatitis A virus (HAV) IRES (Borman and Kean 1997; Ali et al. 2001); this IRES requires the intact eIF4F complex (comprising eIF4E, eIF4A and eIF4G). No viral coding sequences or viral proteins are required for the activity of picornavirus IRES elements.

The FMDV IRES is predicted to have a complex secondary structure which is very similar to that of the EMCV IRES (Pilipenko et al. 1989). These two elements form a discrete class of picornavirus IRES elements which functions very efficiently within the rabbit reticulocyte lysate (RRL) translation system in vitro and within many different cell types (see review by Belsham and Jackson 2000). The entero-/rhinovirus IRES elements form a second class of element which has quite distinct characteristics, e.g. these elements function poorly in the RRL in vitro translation system but their activity is stimulated by the addition of HeLa cell protein extracts. It seems that the HAV IRES (Brown et al. 1994) and the teschovirus IRES (Kaku et al. 2002; Pisarev et al. 2004) elements each have their own unique characteristics which distinguish them from the other two major classes of picornavirus IRES. The sequence identity between the FMDV and EMCV IRES elements is about 50%, but there are some completely identical regions, particularly within the apical region of the I domain and within the J and K domains (see Fig. 2; see also Belsham and Jackson 2000). It is assumed that these highly conserved regions will generally reflect critical regions of the IRES. However, this is not always the case. Within the I domain there is a conserved GNRA tetraloop motif which is important for activity. Tetraloop sequences that fit the GNRA consensus are over-represented, on a statistical basis, within structured RNA elements, and it is believed that they play an important role in RNA-RNA interactions and in RNA-protein interactions (Woese et al. 1990; Costa and Michel 1995). Modification of just the 3′ A residue within this motif greatly diminishes the activity of either the EMCV or FMDV IRES (Roberts and Belsham 1997; Lopez de Quinto and Martinez-Salas 1997; Robertson et al. 1999). The 5′ G residue is also important in the FMDV IRES (Lopez de Quinto and Martinez-Salas 1997). In contrast, the EMCV IRES element seems more tolerant in this respect. EMCV elements containing either a G or an A at this position (i.e. an RNRA motif) display high activity (Robertson et al. 1999), and even an EMCV IRES with a CAGA motif at this position was about 80% as active as the wt element. It is not yet known why this motif is important for IRES activity; no protein interactions have been localised to this site. Direct RNA-RNA interactions between the I domain of the FMDV IRES and other regions of the IRES have been demonstrated (Ramos and Martinez-Salas 1999), and evidence suggests that the structural organization of the IRES is dependent on the GNRA motif (Fernandez-Miragall and Martinez-Salas 2003).

The polypyrimidine tract, located at the 3′ end of the IRES, is an example of a feature which is conserved amongst nearly all picornavirus IRES elements (Meerovitch and Sonenberg 1993) (except for the teschovirus IRES; Kaku et al. 2002) that is less critical for activity. Some mutations within the FMDV polypyrimidine tract are highly detrimental to IRES activity (Kuhn et al. 1990), but surprisingly modification of this tract within the EMCV IRES to a polypurine tract only decreased IRES activity by about 30% (Kaminski et al. 1994). The polypyrimidine tract is usually about 20 nt upstream of an AUG codon. This codon is an initiation site of protein synthesis on the EMCV and FMDV RNAs but not on the entero- or rhinovirus RNAs (translation initiation occurs at the next AUG downstream on these RNAs; see Belsham and Jackson 2000). Indeed on the FMDV RNA the situation is further complicated because initiation of protein synthesis occurs at two sites, 84 nt apart (Sangar et al. 1987; Belsham 1992); the first start site is in the position equivalent to the EMCV start site. The presence of two initiation sites on FMDV RNA is conserved across all seven serotypes of FMDV and is also maintained in the distantly related aphthovirus, ERAV (Hinton et al. 2000). The reason for this arrangement remains unknown.

4 Selection of Initiation Sites of Translation on FMDV RNA

The mechanism of initiation site selection on FMDV and EMCV RNAs has been extensively studied. The initiation codon on EMCV RNA (R-strain) is at nt 834. This is the 11th AUG codon within the viral RNA. There is a strong selectivity for the use of AUG-11 (Kaminski et al. 1990). The AUG-10 codon is just 8 nt upstream but is not used; similarly, usage of downstream AUGs is very low. The first start site on FMDV RNA corresponds, in its position relative to the IRES, to the EMCV AUG-11 but on FMDV RNA many ribosomes fail to initiate at this point and progress to the next AUG codon. The use of two different initiation codons results in the production of two alternate forms of the first component (the Leader protease, L) of the viral polyprotein; these are termed Lab and Lb (see Fig. 1). The context (as defined by Kozak 1989) of the Lab initiation codon is relatively poor compared to that of AUG-11 in the EMCV RNA, and this may partially explain the ability of ribosomes to bypass this site. However, other processes must also be involved because simply improving the context of the Lab start site does not diminish utilisation of the Lb start site (Lopez de Quinto and Martinez-Salas 1999). It is apparent that the FMDV RNA sequence around the two initiation sites has very unusual properties.

Belsham (1992) analysed the utilisation within cells of two different types of mRNA transcript that each contained the two FMDV start sites. One of the expressed RNAs contained the complete FMDV IRES upstream of the two initiation sites, whereas a second mRNA contained only about 60 nt of the 5′-UTR and was translated in a cap-dependent manner. On each transcript, utilisation of both start sites was observed. Furthermore, when two additional AUG codons were introduced in frame between the two FMDV start sites it was shown that all four AUG codons within a sequence of 90 nt were recognised. This occurred independently of whether translation initiation on the transcripts was by the cap-dependent mechanism or was directed by the FMDV IRES. It was concluded that the FMDV IRES directs ribosomes to the Lab AUG initiation codon (as with the EMCV IRES) but many fail to initiate at this point and then scan along the mRNA until another AUG codon is reached. It is remarkable that up to three AUG codons can be bypassed on this stretch of RNA by scanning ribosomes (Belsham 1992).

Subsequent studies showed that the FMDV Lb start site was required for virus infectivity whereas, in contrast, the Lab start site could be removed (Cao et al. 1995; Piccone et al. 1995a). These results prompted further analysis of this system. Studies by Ohlmann and Jackson (1999) indicated that the FMDV IRES is less stringent in its positioning of ribosomes onto the initiation site than the EMCV IRES. For example, with the use of chimaeric RNAs (fused at the polypyrimidine tract) it was found that the EMCV IRES directed greater utilisation of the FMDV Lab start site (corresponding to the EMCV AUG-11 codon) than observed with the FMDV IRES. Furthermore, in a converse experiment, the FMDV IRES generated less efficient selection of AUG-11 on the EMCV RNA sequence than is observed with the EMCV IRES. In other studies, Poyry et al. (2001) showed, with in vitro translation reactions, that the introduction of the iron response element (IRE) between the two FMDV start sites rendered the utilisation of the Lb start site susceptible to inhibition by the binding of the iron response protein (IRP-1) to the IRE. This interaction blocks ribosome scanning but not translation. This result was consistent with the view that the majority of ribosomes reach the Lb start site by scanning, as proposed by Belsham (1992). However, the degree of inhibition observed was less great than observed when the IRE/IRP-1 complex was positioned in the 5′-UTR of a standard mRNA translated by a cap-dependent mechanism. Hence, Poyry et al. (2001) suggested that the results left open the possibility that some ribosomes may reach this initiation site by a different mechanism.

Currently it may be considered that the FMDV IRES directs ribosome attachment to the viral RNA either just upstream or just downstream of the Lab initiation site. Some ribosomes land upstream of the Lab site and can then initiate protein synthesis at this point, but some may fail to do so and then scan along the RNA until the Lb site is reached. The ribosomes which land downstream of the Lab site presumably just migrate along the RNA to initiate translation at the Lb site.

5
Protein Interactions with the FMDV IRES

The FMDV IRES requires essentially all of the canonical translation initiation factors for activity which are needed for the translation of cellular mRNAs (Pilipenko et al. 2000). The only exception is eIF4E (the cap-binding protein) consistent with the lack of a cap structure on the viral RNA. Furthermore, translation of the viral RNA is unimpaired when the translation initiation factor eIF4F is modified by the cleavage of eIF4G (see below). As outlined above, the cap-binding complex eIF4F is a hetero-trimer comprising eIF4E (a protein that binds specifically to the cap structure), eIF4A (an RNA helicase) and eIF4G (a scaffold protein). When eIF4G is cleaved (e.g. as a result of the expression of the FMDV L protease, see below), cap-dependent protein synthesis is inhibited but the C-terminal cleavage product of eIF4G (which retains binding sites for eIF4A and eIF3) is sufficient to support FMDV IRES activity. The N-terminal cleavage product of eIF4G which interacts with eIF4E and the poly(A) binding protein (PABP) is not required for IRES activity (in contrast to initiation on cellular mRNAs). Direct interaction of intact eIF4G (and of the C-terminal cleavage product) with the 3′ region of the FMDV IRES (the J-K domains) has been demonstrated (Lopez de Quinto and Martinez-Salas 2000; Stassinopoulos and Belsham 2001). These results are consistent with data showing that eIF4G protects sequences within the J-K domain of the EMCV IRES from chemical modification (Kolupaeva et al. 1998). There is a good correlation between the ability of IRES elements with mutations within the J-K domain to interact with eIF4G and their activity within cells (Lopez de Quinto and Martinez-Salas 2000). The binding of eIF4G will indirectly bring eIF4A to the IRES, too. The role of eIF4A in IRES function is not clear, but dominant-negative mutants of eIF4A block both cap-dependent translation and picornavirus IRES-directed translation initiation (Pause et al. 1994b; Svitkin et al. 2001a). Early studies on the FMDV IRES also identified a direct interaction between eIF4B and this IRES (Meyer et al. 1995). The binding site for eIF4B has now been mapped to a small stem-loop structure at the extreme 3′ end of the IRES (Lopez de Quinto et al. 2001), but it appears that this interaction has a rather modest effect on the activity of the IRES. This is in agreement with the limited requirement for this factor for the formation of 48S initiation complexes on the EMCV IRES (Pestova et al. 1996).

The FMDV IRES also requires for activity certain cellular proteins in addition to the translation initiation factors. The polypyrimidine tract binding protein (PTB) can be UV cross-linked to all picornavirus IRES elements tested, and multiple binding sites have been mapped on the FMDV and EMCV IRES elements (Kolupaeva et al. 1996). The functional requirement for this interaction varies, however. For example, as judged by in vitro translation assays, the wt EMCV IRES does not need this interaction when linked to its own coding sequence but its activity is stimulated by PTB when other reporter sequences are linked to the IRES (Kaminski and Jackson 1998). The activity of the FMDV IRES is stimulated by PTB in vitro (Niepmann et al. 1997). Furthermore, the stimulation of 48S initiation complex formation on the FMDV IRES with purified components seems to require the interaction of both PTB and ITAF45 (also known as the murine proliferation-associated protein, Mpp1). These two proteins both interact with the same region of the FMDV IRES, and it is not yet clear how they co-operatively stimulate IRES activity (Pilipenko et al. 2000). It is generally believed that the various IRES-interacting proteins, which share the common properties of multiple RNA binding sites and multimeric interactions, stimulate IRES activity by forming or stabilising the tertiary structure of the IRES, and hence they are considered as RNA chaperones (reviewed in Belsham and Sonenberg 1996, 2000).

The poly(rC) binding protein (PCBP2) which stimulates the activity of the PV IRES also binds to the FMDV 5′-UTR (Walter et al. 1999). One site of interaction has been mapped to the domain I of the FMDV IRES (Stassinopoulos and Belsham 2001), but depletion of this protein does not block FMDV IRES activity in vitro (Walter et al. 1999; Stassinopoulos and Belsham 2001). There is a need to establish the role of the various IRES-interacting proteins in the activity of the picornavirus IRES elements within cells, and further cellular proteins that modify IRES function may still need to be identified.

6 The Virus-Encoded Polyprotein

6.1 The L Protease

As indicated above, the first component of the FMDV polyprotein is the Leader (L) protein. FMDV is unique in having a protease as the Leader protein. The L protein is a papain-like cysteine protease and it has at least two distinct activities. It cleaves itself from the rest of the viral polyprotein at the L/P1 junction (Strebel and Beck 1986), and it also induces the cleavage of the translation initiation factor eIF4G (Devaney et al. 1988; Medina et al. 1993); this results in the inhibition of cap-dependent protein synthesis. The cleavage of the L/P1 junction (at a lys/gly bond) can occur *in trans* (Medina et al. 1993) and also probably *in cis* (see Glaser et al. 2002). Each of these functions is shared by the two different forms of the protein, Lab and Lb (Medina et al. 1993). The essential catalytic residues within the protease have been identified (Piccone et al. 1995b: Roberts and Belsham 1995) and the 3-D structure of Lb has been determined (Guarne et al. 1998). A third type of activity displayed by the L protease is the stimulation of enterovirus IRES activity (see Roberts et al. 1998; Hinton et al. 2002). This effect is also achieved by the unrelated entero-/rhinovirus 2A proteases (Hambidge and Sarnow 1992; Borman et al. 1997; Roberts et al. 1998; Sakoda et al. 2001). Both types of viral protease also induce cleavage of eIF4G, but some evidence suggests that IRES activation is a separate process (Hambidge and Sarnow 1992; Roberts et al. 1998; Sakoda et al. 2001) and thus the basis for this effect is not yet understood. It has been suggested that these viral proteases may induce cleavage of another cellular protein which modifies IRES function (Roberts et al. 1998).

The FMDV L protease induces cleavage of eIF4GI very early on within infected cells and requires only very low-level expression of the protease (see Belsham et al. 2000). The cleavage of eIF4GI and the recently identified homologue eIF4GII (Gradi et al. 1998) occurs with similar kinetics within FMDV-infected cells (Gradi et al. 2004). With in vitro assays, direct cleavage of eIF4GI by rather high levels of recombinant FMDV Lb protease has been demonstrated and a cleavage site determined (Kirchweger et al. 1994) which is just seven residues from that cleaved by an enterovirus 2A protease in a similar assay (Lamphear et al. 1993). Similarly, the cleavage cite generated by the L protease in eIF4GII has also been identified in vitro (Gradi et al. 2004). Howerver, there are currently no data on the sites cleaved within eIF4GI or eIF4GII within FMDV-infected cells. It is worth noting that the in vitro cleavage of eIF4GI and eIF4GII by either the FMDV L or entero-/rhinovirus 2A protease is greatly stimulated by the presence of added eIF4E (Haghighat et al. 1996; Ohlmann et al. 1997; Gradi et al. 2004). These results suggest that the conformation of eIF4G is important in determining the sensitivity of the protein to protease digestion and this may, at least in part, explain the difference in levels of protease required for cleavage in vitro and in cells.

The L protease is not essential for FMDV replication; a mutant of the virus lacking the entire Lb coding sequence has been made (Piccone et al. 1995a). The mutant virus replicates in BHK cells (albeit more slowly than the parental virus) but is attenuated in cattle (Brown et al. 1996). The failure to rapidly shut off host cell protein synthesis may permit cells to mount a more efficient anti-viral response (Chinsangaram et al. 1999).

6.2 The Capsid Protein Precursor P1-2A

The FMDV capsid precursor P1-2A is released from the polyprotein by cleavage at its N-terminus by the L protease and at its C-terminus by the 2A protein. Breakage of the viral polyprotein at the 2A/2B junction is very rapid within cells and within in vitro translation reactions, and no uncleaved species are detected. The 2A protein of FMDV is only 18 residues long, and it lacks any protease motifs. The cleavage at the 2A/2B junction only occurs as a co-translational event. The translation of recombinant polyprotein sequences containing mutant forms of the 2A peptide can generate 'uncleaved' proteins; under these circumstances the 2A/2B junction is stable. It also appears that the 2A/2B link is not broken within *E. coli* (Donnelly et al. 1997). Taken together, it seems that the 2A 'cleavage' has to occur on eukaryotic ribosomes during synthesis. It has been proposed that the 2A/2B bond may never normally be made and that the 2A functions by preventing synthesis of the specific peptide bond at the 2A/2B junction (Donnelly et al. 2001).

The P1-2A precursor is processed by the 3C protease to yield VP0 (1AB), VP3 (1C) and VP1 (1D) (see Fig. 1). These are the components of natural empty capsids, and 60 copies of each protein will self-assemble to form particles (Abrams et al. 1995). The stability of these particles is dependent on the myristoylation of the N-terminal glycine residue generated by the L/P1 cleavage (Chow et al. 1987). It is not clear whether empty capsids are a dead-end product or whether they can still participate in the production of new virions. Packaging of the viral RNA to produce virions is associated with the cleavage of the VP0 (1AB) protein into VP4 (1A) and VP2 (1B), but the mechanism of this reaction is not clear.

The FMDV capsid is highly acid labile; it appears that the protonation of His residues, located at the interface of pentameric structures, results in dissociation of the capsid (Curry et al. 1995). This may be the mechanism by which the viral RNA is released from the virus particle after its interaction with integrin receptors on the cell surface and internalisation into endosomes which create a mildly acidic environment (Miller et al. 2001).

6.3
The P2 Precursor

The FMDV 2BC precursor is processed to 2B and 2C by the 3C protease (Fig. 1). Little is known about the function of these proteins. Within PV-infected cells, 2B and 2C are localised within membrane-associated viral replication complexes (Bienz et al. 1987, 1990). Presumably, these proteins also bind to other membranes because the enterovirus 2B protein can increase permeability of cells, as judged by their enhanced susceptibility to the translation inhibitor hygromycin B (van Kuppeveld et al. 1997) and it also blocks protein secretion (Doedens and Kirkegaard 1995). This activity may assist the virus in evading the host immune response. However, some mutants within 2B affect cell growth without apparently affecting either of these functions; thus other activities for the 2B protein may still need to be defined (van Kuppeveld et al. 1997). The various picornavirus 2C protein sequences contain conserved helicase and nucleotide binding motifs, but no evidence for RNA helicase activity has been reported. The 2C protein also determines the sensitivity of viral RNA replication to the inhibitor guanidine (Saunders and King 1982), but the role of 2C in RNA replication is not well defined.

6.4
The P3 Precursor

The FMDV P3 precursor (Fig. 1) is separated from P2 and processed by the 3C protease to 3A, the three distinct copies of the 3B peptide (VPg), 3C protease and 3D RNA polymerase plus a variety of intermediates (e.g. 3CD). The 3A protein has hydrophobic sequences which are believed to anchor it to membranes, and this may be the means by which RNA replication is localised to membrane vesicles. The 3A protein may also serve to deliver the 3B peptides to the sites of RNA replication. Certain strains of FMDV have been isolated that contain in-frame deletions within the 3A coding sequence, and these strains are attenuated in cattle (O'Donnell et al. 2001) but remain highly pathogenic in pigs. This presumably reflects some differences between the two species in the interaction of 3A with cellular factors. It was shown that the expression of the FMDV 3A protein alone disrupts the Golgi apparatus of keratinocytes. It is interesting to note that some picornaviruses, e.g. PV, are greatly inhibited by brefeldin A (Maynell et al. 1992), an agent which modifies vesicle transport between the Golgi and the endoplasmic reticulum. However, in contrast, FMDV is rather insensitive to this agent (O'Donnell et al.

2001), like EMCV (Iruzun et al. 1992). The different sensitivities of the picornaviruses to this agent suggest that the various viruses recruit membranes to their replication complexes in different ways. The PV 3A protein has also been shown to block protein secretion (Doedens and Kirkegaard 1995).

Uniquely, FMDV RNA encodes three distinguishable copies of the 3B peptide. Even ERAV, the only member of the aphthovirus genus other than FMDV, only has a single copy of this peptide. The coding sequences for these different peptides occur as a tandem array in the FMDV genome (see Fig. 1). Each of the FMDV VPgs has been found attached to genomic RNA, and thus each peptide is believed to be functionally equivalent (King et al. 1980). Priming of RNA synthesis requires 3B (VPg). On the basis of results obtained from studies of PV and HRV RNA replication, it is now believed that the initial modification of VPg to VPgpU or VPgpUpU is achieved by using the '*cre*' as a template (Paul et al. 2000; Gerber et al. 2001). The role of the multiple VPg sequences has been investigated by mutagenesis (Falk et al. 1992). Deletion of $3B_3$ alone destroyed virus viability, but this appeared to result from a defect in polyprotein processing rather than a direct effect on RNA replication. It was possible to rescue viable viruses which had a deletion of one or more of the other FMDV 3B sequences, but the mutant viruses replicated less efficiently than the wt virus. It is not yet clear why the presence of three different VPg sequences within the FMDV RNA enhances its replication efficiency (Falk et al. 1992).

6.5 The 3C Protease

The FMDV 3C protease is responsible for most of the proteolytic processing of the viral polyprotein. It functions alone and, in contrast to the PV 3C protease (see Ypma-Wong et al. 1988), does not require 3D sequences for any of its processing activities. The key catalytic residues of the FMDV 3C have been identified (Grubman et al. 1995), and the protease is a member of the trypsin-like family of serine proteases (except that the active serine is replaced by a cysteine). In addition to cleaving the viral polyprotein, the FMDV 3C also modifies certain cellular proteins. The histone H3 was shown to be cleaved by this protease (Falk et al. 1990), and more recently it has been shown that 3C protease also cleaves the translation initiation factors eIF4A and eIF4GI within FMDV-infected cells (Belsham et al. 2000). The precise location of the cleavage site generated by the 3C protease in eIF4AI has been identified

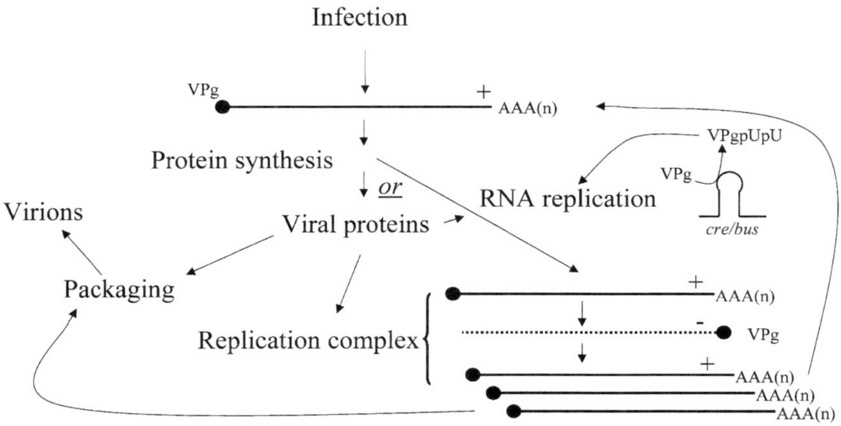

Fig. 3 Functions of viral RNA within cells. The alternative functions of viral RNA within cells are indicated—see text for further details. The initial role of the viral RNA has to be as a mRNA for the production of viral proteins which are required for RNA replication and for packaging of the de novo-synthesised RNA into virions. Some of the different functions of the RNA may occur in discrete compartments within the cell

(Li et al. 2001). The cleavage is specific for eIF4AI, because the closely related eIF4AII (92% identical) is not modified. Two amino acid differences between eIF4AI and eIF4AII near the cleavage site of eIF4AI are sufficient to account for the resistance of eIF4AII to cleavage by 3C. The cleavage of eIF4AI, which will inactivate the protein (Li et al. 2001), may contribute to the decrease in the level of viral protein synthesis during the later phase of virus infection (Belsham et al. 2000). This may facilitate packaging of the viral RNA (see Fig. 3).

The 3C protease cleaves eIF4GI on the C-terminal side of the site generated by the expression of the L protease (Belsham et al. 2000). This activity explains the loss of intact eIF4GI within BHK cells infected by the FMDV mutant lacking the L protease (Piccone et al. 1995a; Belsham et al. 2000). Sequential cleavage of eIF4GI within FMDV-infected BHK cells by the L protease and then of the C-terminal cleavage product of eIF4GI by the 3C protease has been observed; both cleavages are complete by 3 h after infection (G.J. Belsham, unpublished results). Thus at the time of peak viral protein synthesis the form of eIF4GI which supports IRES function is that generated by FMDV 3C. Later on during infection, further modification of eIF4GI occurs and these cleavages may contribute to the decline in viral protein synthesis. The site in eIF4GI cleaved by

FMDV 3C has now been identified (Strong and Belsham 2004) but this is restricted to hamster and murine cells and hence does not seem relevant to the disease process.

6.6 The 3D RNA Polymerase

The 3D protein is the RNA-dependent RNA polymerase. There are two processes required for the replication of picornavirus RNA. Initially, the positive-sense genome acts as a template for the synthesis of an antisense RNA, and this is then used for the production of new positive-sense infectious genomes (see Fig. 3). Within infected cells, a large excess of positive strands over negative strands accumulates. This probably reflects differential recognition of the negative-sense template over the positive-sense template by the RNA polymerase. The 3′ terminus of the positive-sense RNA (the poly(A) tail) and the negative-sense RNA template (antisense S-fragment) are very different in sequence. Thus the nature of the recognition process by the RNA polymerase must be complex but is not currently defined. It is interesting to note that the PV 3D molecule displays co-operativity in its polymerase activity in vitro (Pata et al. 1995) and may usually function as a multimer. The structure of the PV 3D polymerase has been determined by X-ray crystallography, and it is apparent that there is considerable interaction between adjacent molecules in the crystal (Hobson et al. 2001). Whether these properties are relevant to the activity of the protein within an infected cell remains to be determined, but it can be envisaged that a requirement for 3D multimerisation to generate efficient polymerase activity could delay RNA replication until multiple rounds of translation had occurred (see below).

The 3CD molecule does not have RNA polymerase activity, but it does have RNA binding and protease activity. The PV 3CD protein binds to the cloverleaf structure at the 5′ terminus of the PV genome (Gamarnik and Andino 1998), and it is also required for the *cre*-dependent uridylylation of VPg (Paul et al. 2000). No specific role for the FMDV 3CD protein has yet been determined, and as indicated above, FMDV 3C is sufficient for all FMDV polyprotein processing.

7
Structure and Function of the FMDV 3′-UTR

The 3′-UTR of FMDV RNA consists of two components, a region of about 100 nt of heterogeneous sequence and a poly(A) tail. Deletion of the unique (heterogeneous) 3′-UTR sequence blocks infectivity of FMDV RNA (Saiz et al. 2001). This may seem unsurprising but contrasts with the rescue of a viable PV lacking the unique 3′-UTR sequence (Todd et al. 1997). Studies on the 3′-UTR of cardiovirus RNA have shown evidence for three stem-loop structures, one of which is essential for virus viability (Duque and Palmenberg 2001). It is possible that sequences within the viral coding sequence (but perhaps near the 3′ end of the genome) are required to provide the specificity of viral RNA recognition which must be achieved by the replication machinery, especially for the PV mutant lacking the usual 3′-UTR sequences.

It has also been shown that the FMDV 3′-UTR sequence can stimulate the activity of the FMDV IRES (Lopez-de Quinto et al. 2002); furthermore, this effect is independent of the stimulation of IRES activity by poly(A) which was observed previously (Svitkin et al. 2001b). These results suggest possible RNA-RNA interactions between the 5′- and 3′-UTRs or 'bridging' RNA-protein interactions. It is known that PABP [which binds to the poly(A) tail] interacts with eIF4G and with PCBP2. Both of these proteins interact with the FMDV IRES; however, it is not known how these protein-protein interactions would modify IRES activity. Indeed, as indicated above, depletion of PCBP2 has no effect on FMDV IRES activity in vitro but eIF4G is required. Cleavage of eIF4G by the FMDV proteases would disrupt the eIF4G-mediated circularisation of the RNA because PABP binds to the N-terminal region of eIF4G (Imataka et al. 1998) whereas the IRES binds to the C-terminal domain. However, this cleavage also removes competition from the cellular capped mRNAs for the translation machinery, so perhaps a reduced efficiency of translation initiation on the viral RNA could be tolerated under these conditions.

The length of the poly(A) tail determines the infectivity of PV RNA (Spector and Baltimore 1974). This may reflect modification of the stability of the RNA or possibly modification of protein interactions with this sequence; for example, PABP binds optimally to a sequence of at least 25 A residues. As indicated above, interactions between the 5′-UTR and 3′-UTR may influence IRES activity but, furthermore, they may also affect RNA replication. It has been suggested that circularisation of the PV RNA is required for RNA replication and may be achieved through the interaction of the poly(A) binding protein [bound to the 3′ poly(A) tail] with PCBP2 which also binds to the cloverleaf structure at the 5′ end of the genome (Herold and Andino 2001). As mentioned above, no protein interactions with the FMDV S-fragment have yet been identified. However, if the S-fragment does exist in a stable stem-loop structure, as predicted (Clarke et al. 1987; Escarmis et al. 1992), then the 5′ end of the RNA would be brought close to the poly(C) tract which is likely to be bound to PCBP2. Thus PCBP2 could be in close proximity to the 5′ terminus of the FMDV RNA and able to interact with PABP at the 3′ terminus of the genome as proposed for the PV cloverleaf.

8
Interaction Between Translation and Replication—Some Speculation

As indicated in the Introduction, the translation of the viral RNA must precede RNA replication so that viral proteins required for replication are generated within the infected cell. There is an apparent conflict between the use of the viral RNA for translation and the need to replicate the RNA. Studies by Gamarnik and Andino (1998) demonstrated that the passage of ribosomes along the viral RNA prevented RNA replication. Thus it seems that translation of the input RNA must be stopped so that RNA replication can occur (see Fig. 3). The nature of the initial 'switch' between these processes is still unknown. It could be imagined that the mechanism might be common to all picornaviruses and is most easily envisaged as a block to IRES function. Once the 'switch' has occurred it is not necessary for the initial RNA molecule to resume a role in translation because the de novo-synthesised RNAs could perform this function. Thus only a one-way switch is required but clearly it is necessary for translation of viral RNA to continue within the cell. Thus it may be that the processes of RNA replication, RNA packaging into virions and RNA translation are separated into distinct compartments within the cell so that the possibility of competition between these different processes is reduced. However, the in vitro replication of PV (Molla et al. 1992) within a cell-free extract suggests that such compartmentalisation can only be limited at best.

Thus we still have much to learn!

References

Abrams C., King A.M.Q. and Belsham G.J. 1995. Assembly of foot-and-mouth disease virus empty capsids synthesized by a vaccinia virus expression system. J. Gen. Virol. 76:3089–3098

Ali, I.K., McKendrick, L., Morley, S.J. and Jackson, R.J. 2001. Activity of the Hepatitis A virus IRES requires association between the cap-binding translation initiation factor (eIF4E) and eIF4G. J. Virol. 75:7854–7863

Andino, R., Rieckhof, G.E. and Baltimore, D. 1990. A functional ribonucleoprotein complex forms around the 5′ end of poliovirus RNA. Cell 63:369–380

Belsham, G.J. and Bostock, C.J. 1988. Studies on the infectivity of foot-and-mouth disease virus RNA using microinjection. J. Gen. Virol. 69:265–274

Belsham, G.J. and Brangwyn, J.K. 1990. A region of the 5′ non-coding region of foot-and-mouth disease virus RNA directs efficient internal initiation of protein synthesis within cells; interaction with the role of the L protease in translational control. J. Virol. 64:5389–5395

Belsham, G.J. 1992. Dual initiation sites of protein synthesis on foot-and-mouth disease virus RNA are selected following internal entry and scanning of ribosomes in vivo. EMBO J. 11:1105–1110

Belsham, G.J. and Sonenberg, N. 1996. RNA-protein interactions in regulation of picornavirus RNA translation. Microbiol. Rev. 60:499–511

Belsham, G.J. and Jackson, R.J. 2000. Translation initiation on picornavirus RNA. In: 'Translational Control of Gene Expression' Monograph 39. Eds. N. Sonenberg, J.W.B. Hershey and M.B. Mathews. pp869–900. Cold Spring Harbor Laboratory Press, Cold Spring Harbor, NY

Belsham, G.J. and Sonenberg, N. 2000. Picornavirus RNA translation: roles for cellular proteins. Trends Microbiol. 8:330–335

Belsham, G.J., McInerney, G.M. and Ross-Smith, N. 2000. Foot-and-mouth disease virus 3C protease induces cleavage of translation initiation factors eIF4A and eIF4G within infected cells. J. Virol. 74:272–280

Bienz, K., Egger, D., and Pasamontes, L. 1987. Association of polioviral proteins of the P2 genomic region with the viral replication complex and virus-induced membrane synthesis as visualized by electron microscopic immunocytochemistry and autoradiography. Virology 160:220–226

Bienz, K., Egger, D., Troxler, M. and Pasamontes, L. 1990. Structural organization of poliovirus RNA replication is mediated by viral proteins of the P2 genomic region. J. Virol. 64:1156–1163

Borman, A.M., Le Mercier, P., Girard, M. and Kean, K.M. 1997. Comparison of picornaviral IRES-driven internal initiation in cultured cells of different origins. Nucl. Acids Res. 25:925–932

Borman, A.M. and Kean, K.M. 1997. Intact eukaryotic initiation factor 4G is required for hepatitis A virus internal initiation of translation. Virology 237:129–136

Brown, F., Newman, J., Stott, J., Porter, A., Frisby, D., Newton, C., Carey, N. and Fellner, P. 1974. Poly(C) in animal viral RNAs. Nature (London) 251:342–344

Brown, E.A., Zajac, A.J. and Lemon, S.M. 1994. In vitro characterization of an internal ribosomal entry site (IRES) present within the 5′ nontranslated region of hepatitis A virus RNA: comparison with the IRES of encephalomyocarditis virus. J. Virol. 1994 68:1066–1074

Brown, C.C., Piccone, M.E., Mason, P.W., McKenna, T.S.C. and Grubman, M.J. 1996. Pathogenesis of wild-type and leaderless foot-and-mouth disease virus in cattle. J. Virol. 70:5638–5641

Cao, X.M., Bergmann, I.E., Fullkrug, R. and Beck, E. 1995. Functional analysis of the two alternative translation initiation sites of foot-and-mouth-disease virus. J. Virol. 69:560–563

Chinsangaram, J., Piccone, M.E. and Grubman, M.J. 1999. Ability of foot-and-mouth disease virus to form plaques in cell culture is associated with suppression of α/β interferon. J. Virol. 73:9891–9898

Chow, M., Newman, J.F.E., Filman, D., Hogle, J.M., Rowlands, D.J. and Brown, F. 1987. Myristylation of picornavirus capsid protein VP4 and its structural significance. Nature 327:482–486

Clarke, B.E., Brown, A.L., Currey, K.M., Newton, S.E., Rowlands, D.J. and Carroll, A.R. 1987. Potential secondary and tertiary structure in the genomic RNA of FMDV. Nucl. Acids Res. 15:7067–7079

Costa, M. and Michel, F. 1995. Frequent use of the same tertiary motif by self-folding RNAs. EMBO J. 14:1276–1285

Crawford, N.M. and Baltimore, D. 1983. Genome-linked protein VPg of poliovirus is present as free VPg and VPgpUpU in poliovirus-infected cells. Proc. Natl. Acad. Sci. USA. 80:7452–7455

Curry, S., Abrams, C.C., Fry, E., Crowther, J.C., Belsham, G.J., Stuart, D.I. and King, A.M.Q. 1995. Viral RNA modulates the acid sensitivity of foot-and-mouth disease virus capsids. J. Virol. 69:430–438

Devaney, M.A., Vakharia, V.N., Lloyd, R.E., Ehrenfeld, E. and Grubman, M.J. 1988. Leader protein of foot-and-mouth-disease virus is required for cleavage of the p220 component of the cap-binding protein complex. J. Virol. 62:4407–4409

Doedens, J.R. and Kirkegaard, K. 1995. Inhibition of cellular protein secretion by poliovirus proteins 2B and 3A. EMBO J. 14:894–907

Donnelly, M.L., Gani, D., Flint, M., Monaghan, S. and Ryan, M.D. 1997. The cleavage activities of aphthovirus and cardiovirus 2A proteins. J. Gen. Virol. 78:13–21

Donnelly, M.L.L., Luke, G., Mehrotra, A., Li, X., Hughes, L.E., Gani, G. and Ryan, M.D. 2001. Analysis of the aphthovirus 2A/2B polyprotein 'cleavage' mechanism indicates not a proteolytic reaction, but a novel translational effect: a putative ribosomal 'skip'. J. Gen. Virol. 82:1013–1025

Duke, G.M., Osorio, J.E. and Palmenberg, A.C. 1990. Attenuation of mengo-virus through genetic-engineering of the 5′ noncoding poly(C) tract. Nature 343:474–476

Duque, H. and Palmenberg, A.C. 2001. Phenotypic characterization of three phylogenetically conserved stem-loop motifs in the mengovirus 3′ untranslated region. J. Virol. 75:3111–3120

Escarmis, C., Toja, M., Medina, M. and Domingo, E. 1992. Modifications of the 5′ untranslated region of foot-and-mouth disease virus after prolonged persistence in cell culture. Virus Res. 26:113–125

Escarmis, C., Dopazo, J., Davila, M., Palma, E.L. and Domingo, E. 1995. Large deletions in the 5′-untranslated region of foot-and-mouth-disease virus of serotype-C. Virus Res. 35:155–167

Falk, M.M., Grigera, P.R., Bergmann, I.E., Zibert, A., Multhaup, G. and Beck, E. 1990. Foot-and-mouth-disease virus protease-3C induces specific proteolytic cleavage of host-cell histone-H3. J. Virol. 64:748–756

Fernandez-Miragall, O. and Martinez-Salas, F. 2003. Structured organization of a viral IRES depends on the integrity of the GNRA motif. RNA 9:1333–1344

Falk, M.M., Sobrino, F. and Beck, E. 1992. VPg-gene amplification correlates with infective particle formation in foot-and-mouth-disease virus J. Virol. 66:2251–2260

Gamarnik, A.V. and Andino, R.. 1997. Two functional complexes formed by KH domain containing proteins with the 5′ noncoding region of poliovirus RNA. RNA 3:882–892

Gamarnik, A.V. and Andino, R. 1998. Switch from translation to RNA replication in a positive-stranded RNA virus. Genes Dev. 12:2293–2304

Gerber, K., Wimmer, E. and Paul, A.V. 2001. Biochemical and genetic studies of the initiation of human rhinovirus 2 RNA replication: identification of a *cis*-replicating element in the coding sequence of 2Apro. J Virol. 75:10979–10990

Gingras, A.C., Raught, B. and Sonenberg, N. 1999. eIF4 initiation factors: effectors of mRNA recruitment to ribosomes and regulators of translation. Annu. Rev. Biochem. 68:913–963

Glaser, W., Cencic, R. and Skern, T. 2001. Foot-and-mouth disease virus Leader proteinase: involvement of C-terminal residues in self-processing and cleavage of eIF4GI. J. Biol. Chem. 276:35473–35481

Goodfellow, I., Chaudhry, Y., Richardson, A., Meredith, J., Almond, J.W., Barclay, W. and Evans, D.J. 2000. Identification of a *cis*-acting replication element within the poliovirus coding region. J. Virol. 74:4590–4600

Gradi, A., Imataka, H., Svitkin, Y.V., Rom, E., Raught, B., Morino, S. and Sonenberg, N. 1998. A novel functional human eukaryotic translation initiation factor 4G. Mol. Cell. Biol. 18:334–342

Gradi, A., Foeger, N., Strong, R., Svitkin, Y.V., Sonenberg, N., Skern, T., and Belsham, G.J. 2004. Cleavage of enkaryotic translation initiation factor 4GII within foot-and-mouth disease virus-effected cells: identification of the L-protease cleavage site in vitro. J. Virol. 78:3271–3278

Grubman, M.J., Zellner, M., Bablanian, G., Mason, P.W. and Piccone, M.E. 1995. Identification of the active-site residues of the 3C proteinase of foot-and-mouth-disease virus. Virology 213:581–589

Guarne, A., Tormo, J., Kirchweger, R., Pfistermueller, D., Fita, I. and Skern, T. 1998. Structure of the foot-and-mouth disease virus leader protease: a papain-like fold adapted for self-processing and eIF4G recognition. EMBO J. 17:7469–7479

Haghighat, A., Svitkin, Y., Novoa, I., Kuechler, E., Skern, T. and Sonenberg, N. 1996. The eIF4G-eIF4E complex is the target for direct cleavage by the rhinovirus 2A proteinase. J. Virol. 70:8444–8450

Hahn, H. and Palmenberg. A.C. 1995. Encephalomyocarditis viruses with short poly(C) tracts are more virulent than their mengovirus counterparts. J. Virol. 69:2697–2699

Hambidge, S.J. and Sarnow, P. 1992. Translational enhancement of the poliovirus 5′ noncoding region mediated by virus-encoded polypeptide 2A. Proc. Natl. Acad. Sci. USA. 89:10272–10276

Harris, T.J.R. and Brown, F. 1977. Biochemical analysis of a virulent and an avirulent strain of foot-and-mouth disease virus. J. Gen. Virol. 34:87–105

Herold, J. and Andino, R. 2001. Poliovirus RNA replication requires genome circularization through a protein-protein bridge. Mol. Cell 7:581–591

Hinton, T.M., Li, F. and Crabb, B.S. 2000. Internal ribosomal entry site-mediated translation initiation in equine rhinitis A virus: similarities to and differences from that of foot-and-mouth disease virus. J. Virol. 74:11708–11716

Hinton, T., Ross-Smith, N., Warner, S., Belsham, G.J. and Crabb, B. 2002. Conservation of L and 3C proteinase activities across distantly related aphthoviruses. J. Gen. Virol. 83:3111–3121

Hobson, S.D., Rosenblum, E.S., Richards, O.C., Richmond, K., Kirkegaard, K., and Schultz S.C. 2001. Oligomeric structures of poliovirus polymerase are important for function. EMBO J 20:1153–1163

Imataka, H., Gradi, A. and Sonenberg N. 1998. A newly identified N-terminal amino acid sequence of human eIF4G binds poly(A)-binding protein and functions in poly(A)-dependent translation. EMBO J. 17:7480–7489

Irurzun, A., Perez, L. and Carrasco, L. 1992. Involvement of membrane traffic in the replication of poliovirus genomes: effects of brefeldin A. Virology 191:166–175

Jang, S.K., Krausslich, H.G., Nicklin, M.J., Duke, G.M., Palmenberg, A.C. and Wimmer, E. 1998. A segment of the 5′ nontranslated region of encephalomyocarditis virus RNA directs internal entry of ribosomes during in vitro translation. J. Virol. 62:2636–2643

Kaku, Y., Chard, L.S., Inoue, T. and Belsham, G.J. 2002. Unique characteristics of a picornavirus internal ribosome entry site from the Porcine Teschovirus-1 Talfan. J. Virol. 76:11721–11728

Kaminski, A., Howell, M.T. and Jackson, R.J. 1990. Initiation of encephalomyocarditis virus RNA translation: the authentic initiation site is not selected by a scanning mechanism. EMBO J. 9:3753–3759

Kaminski, A., Belsham, G.J. and Jackson, R.J. 1994. Translation of encephalomyocarditis virus RNA: parameters influencing the selection of the internal initiation site. EMBO J. 13:1673–1681

Kaminski, A. and Jackson, R.J. 1998. The polypyrimidine tract binding protein (PTB) requirement for internal initiation of translation of cardiovirus RNAs is conditional rather than absolute. RNA 4:626–638

King, A.M.Q., Sangar, D.V., Harris, T.J.R. and Brown F. 1980. Heterogeneity of the genome-linked protein of FMDV. J. Virol. 34:627–634

Kirchweger, R., Ziegler, E., Lamphear, B.J., Waters, D., Liebig, H.D., Sommergruber, W., Sobrino, F., Hohenadl, C., Blaas, D., Rhoads, R.E. and Skern, T. 1994. Foot-and-mouth disease virus leader proteinase: purification of the Lb form and determination of its cleavage site on eIF-4γ. J. Virol. 61:2711–2718

Kolupaeva, V.G., Hellen, C.U.T. and Shatsky, I.N. 1996. Structural analysis of the interaction of the pyrimidine tract-binding protein with the internal ribosomal entry site of encephalomyocarditis virus and foot-and-mouth disease virus RNAs. RNA 2:1199–1212

Kolupaeva, V.G., Pestova, T.V., Hellen, C.U.T. and Shatsky I.N. 1998. Translation eukaryotic initiation factor 4G recognizes a specific structural element within the internal ribosome entry site of encephalomyocarditis virus RNA. J. Biol. Chem. 273:18599-18604

Kozak, M. 1989. The scanning model for translation: an update. J. Cell Biol. 108:229–241

Kuhn, R., Luz, N. and Beck, E. 1990. Functional analysis of the internal translation initiation site of foot-and-mouth disease virus. J. Virol. 64:4625–4631

Lamphear, B.J., Yan, R.Q., Yang, F., Waters, D., Liebig, H.D., Klump, H., Kuechler, E., Skern, T. and R.E. Rhoads. 1993. Mapping the cleavage site in protein synthesis initiation factor eIF-4γ of the 2A proteases from human coxsackievirus and rhinovirus. J. Biol. Chem. 268:19200–19203

Li, W., Ross-Smith, N., Proud, C.G. and Belsham, G.J. 2001. Cleavage of translation initiation factor 4AI (eIF4AI) but not eIF4AII by foot-and-mouth disease virus 3C protease: determination of the eIF4AI cleavage site. FEBS Lett. 507:1–5

López de Quinto, S. and Martínez-Salas, E. 1997. Conserved structural motifs located in distal loops of aphthovirus internal ribosome entry site domain 3 are required for internal initiation of translation. J. Virol. 71:4171–4175

López de Quinto, S. and Martinez-Salas, E. 1999. Involvement of the aphthovirus RNA region located between the two functional AUGs in start codon selection. Virology 255:324–336

López de Quinto, S. and Martínez-Salas, E. 2000. Interaction of the eIF4G initiation factor with the aphthovirus IRES is essential for internal translation initiation in vivo. RNA 6:1380–1392

López de Quinto, S., Lafuente, E. and Martínez-Salas, E. 2001. IRES interaction with translation initiation factors: functional characterization of novel RNA contacts with eIF3, eIF4B, and eIF4GII. RNA 7:1213–1226

López de Quinto, S., Saiz, M., de la Morena, D., Sobrino, F. and Martínez-Salas, E. 2002. IRES-driven translation is stimulated separately by the FMDV 3′ NCR and poly(A) sequences. Nucl. Acids Res. 30:4398–4405

Martin, L.R. and Palmenberg, A.C. 1996. Tandem mengovirus 5′ pseudoknots are linked to viral RNA synthesis, not poly(C)-mediated virulence. J. Virol. 70:8182–8186

Mason, P.W., Bezborodova, S.V. and Henry, T.M. 2002. Identification and characterization of a *cis*-acting replication element (cre) adjacent to the IRES of foot-and-mouth disease virus. J. Virol. 76:9686–9694

Maynell, L. A., Kirkegaard, K. and Klymkowsky, M.W. 1992. Inhibition of poliovirus RNA synthesis by brefeldin A. J. Virol. 66:1985–1994

McKnight, K.L. and Lemon.S.M. 1996. Capsid coding sequence is required for efficient replication of human rhinovirus-14 RNA. J. Virol. 70:1941–1952

McKnight, K. L. and Lemon, S.M. 1998. The rhinovirus type 14 genome contains and internally located RNA structure that is required for viral replication. RNA 4:1569–1584

Medina, M., Domingo, E., Brangwyn, J.K. and Belsham, G.J. 1993. The two species of the foot-and-mouth disease virus leader protein, expressed individually, exhibit the same activities. Virology 194:355–359

Meerovitch, K. and Sonenberg, N. 1993. Internal initiation of picornavirus RNA translation. Semin. Virol. 4:217–227

Meyer, K., Petersen, A., Niepmann, M. and Beck, E. 1995. Interaction of eukaryotic initiation factor eIF-4B with a picornavirus internal translation initiation site. J. Virol. 69:2819–2824

Miller, L.C., Blakemore, W., Sheppard, D., Atakilit, A., King, A.M.Q. and Jackson, T. 2001. Role of the cytoplasmic domain of the β-subunit of integrin $\alpha v\beta 6$ in infection by foot-and-mouth disease virus. J. Virol. 75:4158–4164

Molla, A., Paul, A.V. and Wimmer E. 1991. Cell-free, de novo synthesis of poliovirus. Science 254:1647–1651

Murray, K.E., Roberts, A.W. and Barton, D.J. 2001. Poly(rC) binding proteins mediate poliovirus mRNA stability. RNA 7:1126–1141

Niepmann, M., Petersen, A., Meyer, K. and Beck, E. 1997. Functional involvement of polypyrimidine tract-binding protein in translation initiation complexes with the internal ribosome entry site of foot-and-mouth disease virus. J. Virol. 71:8330–8339

O'Donnell, V.K., Pacheco, J.M., Henry, T.M. and Mason, P.W. 2001. Subcellular distribution of the foot-and-mouth disease virus 3A protein in cells infected with viruses encoding wild-type and bovine-attenuated forms of 3A. Virology 287:151–162

Ohlmann, T., Pain, V.M., Wood, W., Rau, M. and Morley, S.J. 1997. The proteolytic cleavage of eukaryotic initiation factor (eIF) 4G is prevented by eIF4E binding protein (PHAS-I; 4E-BP1) in the reticulocyte lysate. EMBO J. 16:844–855

Ohlmann T. and Jackson, R.J. 1999. The properties of chimeric picornavirus IRESes show that discrimination between internal translation initiation sites is influenced by the identity of the IRES and not just the context of the AUG codon. RNA 5:764–778

Parsley, T.B., Towner, J.S., Blyn, L.B., Ehrenfeld, E. and Semler, BL. 1997. Poly (rC) binding protein 2 forms a ternary complex with the 5′-terminal sequences of poliovirus RNA and the viral 3CD proteinase. RNA 3:1124–1134

Pata, J.D., Schultz, S.C. and Kirkegaard, K. 1995. Functional oligomerization of poliovirus RNA-dependent RNA-polymerase. RNA 1:466–477

Paul, A.V., Rieder, E., Kim, D.W., van Boom, J.H. and Wimmer, E. 2000. Identification of an RNA hairpin in poliovirus RNA that serves as the primary template in the in vitro uridylylation of VPg. J. Virol. 74:10359–10370

Pause, A., Belsham, G.J., Gingras, A-C., Donze, O., Lin, T-A., Lawrence, J.C. Jr. and Sonenberg, N. 1994a. Insulin dependent stimulation of protein synthesis by phosphorylation of a novel regulator of 5′-cap function. Nature (London) 371:762–767

Pause, A., Methot, N., Svitkin, Y.V., Merrick, W.C. and Sonenberg. 1994b. Dominant negative mutants of mammalian translation initiation factor eIF-4A define a critical role for eIF-4F in cap-dependent and cap-independent initiation of translation. EMBO J. 13:1205–1215

Pelletier, J., and Sonenberg, N. 1998. Internal initiation of translation of eukaryotic mRNA directed by a sequence derived from poliovirus RNA. Nature (London) 334:320–325

Pestova, T.V., Hellen, C.U.T. and Shatsky, I.N. 1996. Canonical eukaryotic initiation factors determine initiation of translation by internal ribosomal entry. Mol. Cell. Biol. 16:6859–6869

Piccone, M.E., Rieder, E., Mason, P.W. and Grubman, M.J. 1995a. The foot-and-mouth-disease virus leader proteinase gene is not required for viral replication. J. Virol. 69:5376–5382

Piccone, M.E., Zellner, M., Kumosinski, T.F., Mason, P.W. and Grubman, M.J. 1995b. Identification of the active-site residues of the L-proteinase of foot-and-mouth-disease virus. J. Virol. 69:4950–4956

Pilipenko, E.V., Blinov, V.M., Chernov, B.K., Dmitrieva, T.M. and Agol, V.I. 1989. Conservation of the secondary structure elements of the 5′′-untranslated region of cardiovirus and aphthovirus RNAs. Nucl. Acids Res.17:5701–5711

Pilipenko, E.V., Pestova, T.V., Kolupaeva, V.G., Khitrina, E.V., Poperechnaya, A.N., Agol, V.I. and Hellen, C.U.T. 2000. A cell cycle-dependent protein serves as a template-specific translation initiation factor. Gen. Dev. 14:2028–2045

Pisarev,. A.Y., Charch, L.S., Kaku, Y., Johns, H.L., Shatsky, I.N., and Belsham, G.J. 2004. Functional and structural similarities between the internal ribosome entry sites of hepatitis C virus and porcine teschovirus, a picornavirus. J. Virol. 78:4487–4497

Poyry, T.A.A., Hentze, M.W. and Jackson, R.J. 2001. Construction of regulatable picornavirus IRESes as a test of current models of the mechanism of internal translation initiation. RNA 7:647–660

Ramos, R. and Martínez-Salas, E. 1999. Long-range RNA interactions between structural domains of the aphthovirus internal ribosome entry site (IRES). RNA 5:1374–1383

Rieder, E., Bunch, T., Brown, F. and Mason, P.W. 1993. Genetically-engineered foot-and-mouth-disease viruses with poly(C) tracts of 2 nucleotides are virulent in mice. J. Virol. 67:5139–5145

Roberts, P. and Belsham, G.J. 1995. Identification of critical amino acids within the foot-and-mouth disease virus Leader protein, a cysteine protease. Virology 213:140–146

Roberts, L.O. and Belsham G.J. 1997. Complementation of defective picornavirus internal ribosome entry site (IRES) elements by the coexpression of fragments of the IRES. Virology 227:53–62

Roberts, L.O., Seamons, R.A. and Belsham, G.J. 1998. Recognition of picornavirus internal ribosome entry sites within cells; influence of cellular and viral proteins. RNA 4:520–529

Robertson, M.E., Seamons, R.A. and Belsham, G.J. 1999. A selection system for functional internal ribosome entry site (IRES) elements: analysis of the requirement for a conserved GNRA tetraloop in the encephalomyocarditis virus IRES. RNA 5:1167–1179

Saiz, M., Gomez, S., Martinez-Salas, E. and Sobrino, F. 2001. Deletion or substitution of the aphthovirus 3′ NCR abrogates infectivity and virus replication. J. Gen. Virol. 82:93–101

Sakoda, Y., Ross-Smith, N., Inoue, T. and Belsham, G.J. 2001. An attenuating mutation in the 2A protease of swine vesicular disease virus, a picornavirus, regulates cap- and internal ribosome entry site-dependent protein synthesis. J Virol. 75, 10643–10650

Sangar, D.V., Newton, S.E., Rowlands, D.J. and Clarke, B.E. 1987. All foot and mouth disease serotypes initiate protein synthesis at two separate AUGs. Nucl. Acids Res. 15:3305–3315

Saunders, K. and King, A.M.Q. 1982. Guanidine-resistant mutants of aphthovirus induce the synthesis of an altered non-structural polypeptide, p34. J. Virol. 42:389–394

Spector, D.H. and Baltimore, D. 1974. Requirement of 3′ terminal polyadenylic acid for the infectivity of poliovirus RNA. Proc. Natl. Acad. Sci. USA 71:2983–2987

Stassinopoulos, I.A. and Belsham, G.J. 2001. A novel protein-RNA binding assay: functional interactions of the foot-and-mouth disease virus internal ribosome entry site with cellular proteins. RNA 7:114–122

Strebel, K., and Beck, E. 1986. A second protease of foot-and-mouth-disease virus. J. Virol. 58:893–899

Strong, R., and Belsham, G.J. 2004. Sequential modification of translation initiation factor elF4GI by two different foot-and-mouth disease protease within infected baby hamster kidney cells: identification of the $3e^{pro}$ cleavage site. J. Gen. Virol. 85 (in press)

Svitkin, Y.V., Pause, A., Haghighat, A., Pyronnet, S., Witherell, G., Belsham, G.J. and Sonenberg, N. 2001a. The requirement for eukaryotic initiation factor 4A (eIF4A) in translation is directly proportional to the degree of mRNA 5' secondary structure. RNA 7:382–394

Svitkin, Y.V., Imataka, H., Khaleghpour, K., Kahvejian, A., Liebig, H.D. and Sonenberg, N. 2001b. Poly(A)-binding protein interaction with eIF4G stimulates picornavirus IRES-dependent translation. RNA 7:1743–52

Tiley, L., King, A.M.Q. and Belsham, G.J. 2003. The foot-and-mouth disease virus *cis*-acting replication element (cre) can be complemented *in trans* within infected cells. J. Virol. (in press)

Todd, S., Towner, J.S., Brown, D.M. and Semler, B.L. 1997. Replication-competent picornaviruses with complete genomic RNA 3′ noncoding region deletions. J. Virol. 71:8868–8874

van Kuppeveld, F.J.M., Melchers, W.J.G, Kirkegaard, K. and Doedens, J.R. 1997. Structure-function analysis of coxsackie B3 virus protein 2B. Virology 227:111–118

Walter, B.L., Nguyen, J.H., Ehrenfeld, E. and Semler, B.L. 1999. Differential utilization of poly(rC) binding protein 2 in translation directed by picornavirus IRES elements. RNA 5:1570–1585

Woese, C.R., Winker, S. and Gutell, R.R. 1990. Architecture of ribosomal-RNA—constraints on the sequence of tetra-loops. Proc. Natl. Acad. Sci. USA 87:8467–8471

Ypma-Wong, M.F., Dewalt, P.G., Johnson, V.H., Lamb, J.G., and Semler, B.L. 1988. Protein 3CD is the major poliovirus proteinase responsible for the cleavage of the P1 capsid precursor. Virology 166:265–270

CTMI (2005) 288:71–101

The Structure of Foot-and-Mouth Disease Virus

E. E. Fry[1] · D. I. Stuart[1] (✉) · D. J. Rowlands[2]

[1] Division of Structural Biology, Wellcome Trust Centre for Human Genetics, Roosevelt Drive, Oxford, OX3 7BN, UK
enquiries@strubi.ox.ac.uk
[2] School of Biochemistry and Microbiology, University of Leeds, Woodhouse Lane, Leeds, LS2 9JT, UK

Abstract Structural studies of foot-and-mouth disease virus (FMDV) have largely focused on the mature viral particle, providing atomic resolution images of the spherical protein capsid for a number of sero- and sub-types, structures of the highly immunogenic surface loop, Fab and GAG receptor complexes. Additionally, structures are available for a few non-structural proteins. The chapter reviews our current structural knowledge and its impact on our understanding of the virus life cycle proceeding from the mature virus through immune evasion/inactivation, cell-receptor binding and replication and alludes to future structural targets.

1 Introduction

The ultimate goal of structural studies of foot-and-mouth disease virus (FMDV) is to visualize the virus, its sub-components and its interactions with the host cell at a molecular level. The progress to date is summarised in Fig. 1. In essence, FMDV targets epithelial cells and is internalised after interaction with an integrin receptor. Within the endosome, a reduction in pH induces disruption of the capsid and ejection of the RNA genome into the cytoplasm, where it is translated as a single open reading frame. The resulting polypeptide undergoes a series of cleavages to produce in excess of 12 mature polyprotein and partial cleavage intermediates.

Many of the non-structural proteins and precursors, as well as the RNA genome, make up the ribonucleoprotein replication complex which forms on the surface of membranous vesicles present in infected cells. This complex copies the positive-strand viral RNA into negative-strand complementary RNA which then functions as template for the production of multiple positive RNA strands to be used for translation, replication and packaging into viral progeny.

This chapter describes the work behind the structures depicted in Fig. 1, introduced from a historical perspective and proceeding from the mature virus through immune evasion/inactivation, cell-receptor binding and on to replication, alluding to future structural targets along the way.

FMDV was the first animal virus to be identified as a filterable agent (Loeffler and Frosch 1897), although at the time it could not be visual-

Fig. 1 A structural view of the viral life cycle. The background tomography of a cell is taken from Ohad et al. (2002). Membrane sections are taken from the X-ray crystallographic structure of the PRD1 bacteriophage (Cockburn et al., in preparation). The RNA is constructed from coordinates deposited as 157D (Berman et al. 2000). Host components are *grey*, and viral components are coloured. The figure shows the all-atom structure of reduced O serotype virus [structure determined in the presence of DTT (Logan et al. 1993)]. Atoms are depicted as *solid spheres* with radii corresponding to their size (*CPK*): VP1 is *blue*, VP2 *green* and VP3 *red* (VP4 is internal and not visible in these views). Individual viral proteins are depicted as above. The integrin $\alpha v\beta 3$ cellular receptor (1L5G, Berman et al. 2000) is depicted in *dark* and *mid-grey*. An inactivated C-serotype virus complexed with Fab is shown colour coded as above with the Fab in *grey* (1QGC, Berman et al. 2000). Cryo-EM representa-

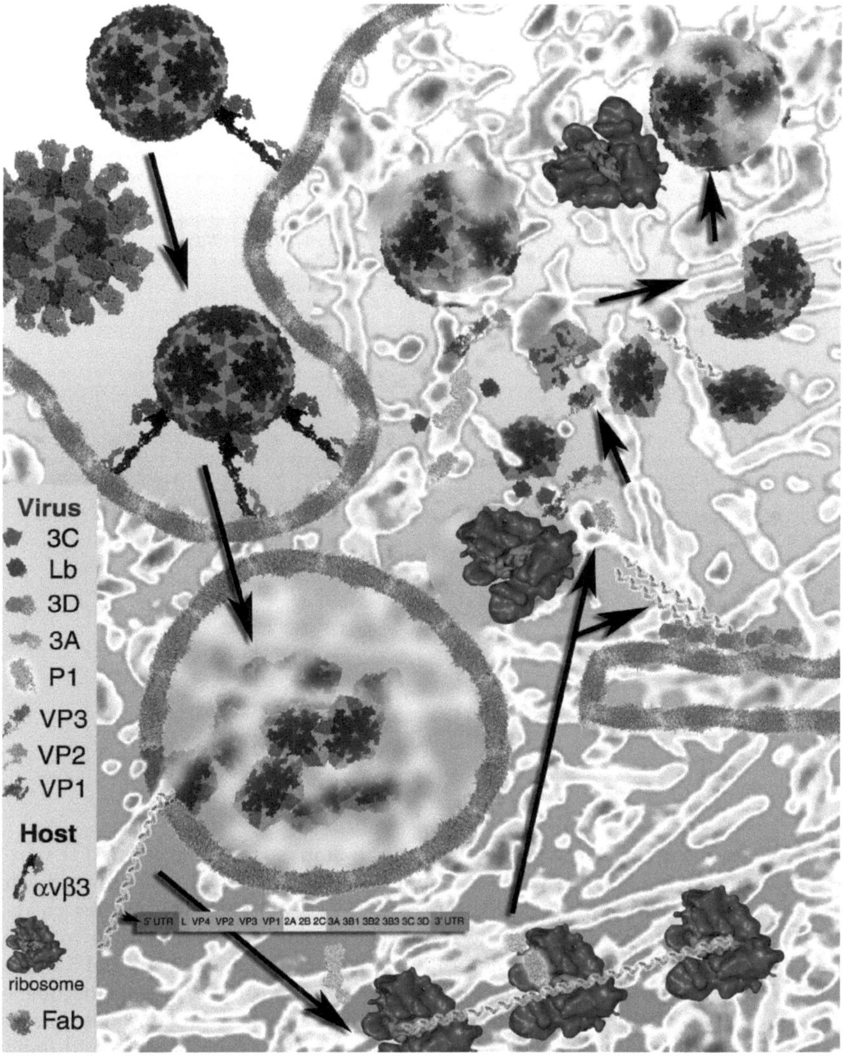

tions of an active ribosome are depicted in *grey* with difference density interpreted as a pseudoknot and tRNAs shown in *green* (Spanchak et al., in preparation). Nascent proteins are depicted in *yellow.* Poliovirus 3D polymerase (1RDR, Berman et al. 2000) is depicted in *rose.* The poliovirus 3C proteinase (1L1N, Berman et al. 2000) is depicted in *magenta* and the FMDV leader proteinase dimer in *purple* and *orange.* The dimeric N-terminal fragment of poliovirus 3A (1NG7, Berman et al. 2000) is shown in *pink* and *green.* All molecular renditions were prepared with Bobscript (Esnouf 1997) and Raster3D (Merritt and Murphy 1994)

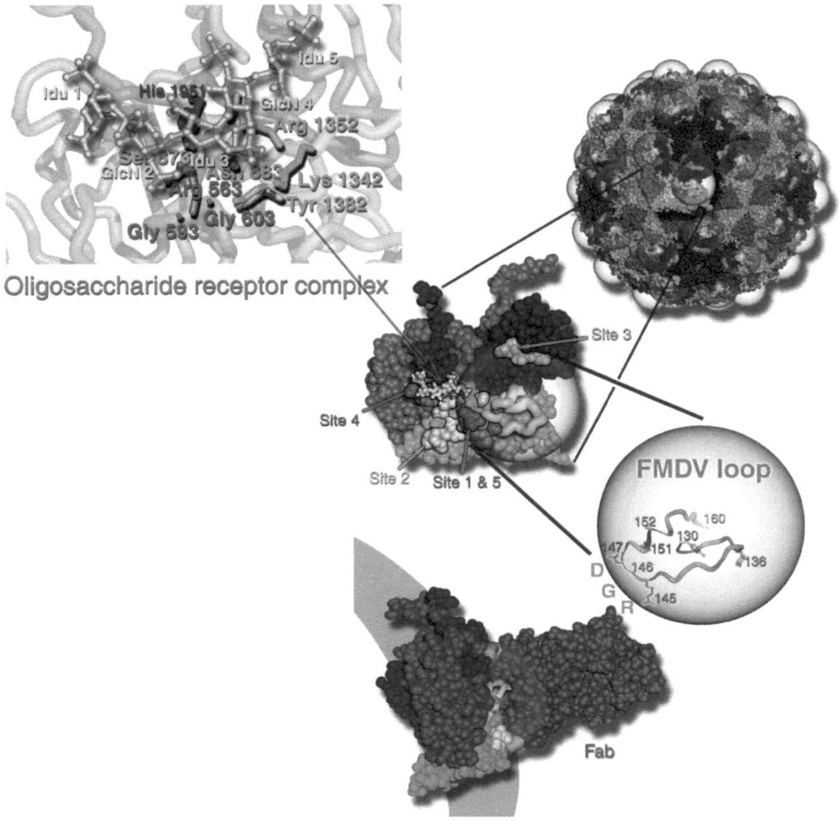

Fig. 2 Virus interactions. The figure shows the all-atom structure of reduced O serotype virus (as in Fig. 1). The VP1 GH loop residues 130–160 (FMDV loop) is shown as a 'worm' in *cyan* with the Arg-Gly-Asp residues in *orange* CPK (speres corresponding to atomic radii). Antigenic residues are colour-coded according to their classification into sites (Crowther et al. 1993; Kitson et al. 1990): sites 1 and 5 (*mid-blue*), site 2 (*pale yellow*), site 3 (*light blue*), site 4 (*magenta*). The potential occupancy of the VP1 GH loop [which appears to behave as a mobile domain hinged at its base (Parry et al. 1990)] is modelled by a *transparent sphere* centred at the mid-point between the two ends of the loop. The heparin motif which has been visualized in complex with the virus (Fry et al. 1999) is drawn in *yellow ball-and-stick*. A protomeric subunit (the smallest repeating unit, representing 1/60th of the capsid) is outlined and enlarged. The FMDV loop is further enlarged with residues that are conserved (or virtually conserved) across all strains drawn in full. In this depiction of the loop sequence variability between strains is proportional to the thickness of the tube; thus the most variable regions are thicker and the most conserved regions thinner. The residue numbers are shown for the conserved residues with the residue type being indicated for the proposed integrin binding RGD motif (*orange*). All resi-

ised. Elucidation of its general morphology was not possible until the advent of the electron microscope, when negative-stained images to a resolution of 40–50 Å revealed smooth, round particles of 30-nm diameter (Wild et al. 1969). A detailed understanding of its structure came two decades later when advances in technology and the production of crystals of the virus (Fox et al. 1987) permitted the application of X-ray crystallography. The X-ray diffraction data, together with knowledge of the complete protein sequence and structural information from related picornaviruses, led to the atomic resolution structure for the protein shell of an O serotype virus (Acharya et al. 1989) (Figs. 1, 2). Subsequent studies permitted comparisons with other serotypes and subtypes, e.g. the European A, O and C serotypes (Fig. 3) (Acharya et al. 1989; Curry et al. 1996; Lea et al. 1995; Lea et al. 1994), and monoclonal antibody-resistant mutants (MAR mutants) (Parry et al. 1990). A striking feature of all of these virus particle structures was that a particularly interesting sequence within the VP1 protein, which had been shown to be of major antigenic and immunogenic importance, was not visible because of its disordered state relative to the remainder of the particle surface. Contemporaneously with the structural work this poorly visualised surface loop was also identified as being important in interactions with host cell receptors. In addition, the MAR mutant results suggested, unexpectedly, that a single mutation might produce significant structural rearrangements in this sequence. Often referred to as the FMDV loop, this structure has since been more clearly visualised (Logan et al. 1993) (Fig. 2) and its role in antibody and receptor recognition delineated with some precision. It appears from these studies that neutralisation of FMDV occurs by blocking of receptor attachment. Indeed, complexes of the virus with large fragments of monoclonal antibodies (Fabs) have been visual-

dues are numbered such that the least significant digit defines the protein chain, e.g. 1382 is residue 138 of VP2. The heparin binding site is enlarged to show the protein side chains which act as ligands for the five sugar residues IDU1–IDU5. The protein backbone is shown in the background colour-coded as above and the side-chains interacting with the heparin are in ball-and-stick and coloured as the protein backbone; His195 of VP1, Lys134, Arg135 and Tyr138 of VP2 and Arg56, Gly59, Gly60, Ser87 and Asn88 of VP3. A Fab fragment interacting with an FMDV C serotype protomeric subunit is viewed from the side. This model was produced from a cryo-EM reconstruction (1QGC, Berman et al. 2000). The virus is depicted as above with theVP1 GH loop in *cyan*, the RGD motif in *orange ball-and-stick* and antigenic residues highlighted. The Cαs of the Fab fragment are depicted in *grey* CPK

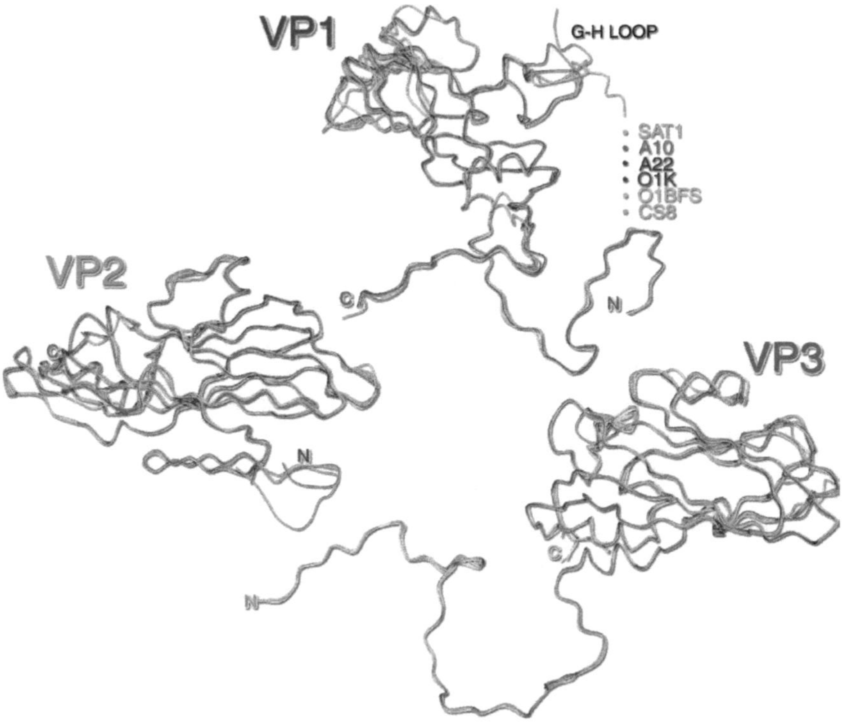

Fig. 3 Structural comparison of serotypes. A superimposition of the VP1, 2 and 3 proteins of FMDVs: A10 (E.E. Fry. et al., unpublished data), A22 (Curry et al. 1996), O1BFS (Acharya et. 1989), O1K (Lea et al. 1995), CS8 (Lea et al. 1994) and SAT1 (Adams et al., unpublished data). Each is drawn as a thin worm coloured according to the key. The O serotype structures depicted are not reduced; hence the VP1 G-H loop is not shown

ised by cryoelectron microscopy (Figs. 1, 2) (Hewat et al. 1997) and synthetic peptides representing antigenic portions of the FMDV loop have been imaged complexed with Fabs by X-ray crystallography (Verdaguer et al. 1994, 1999).

The identification of cellular receptors has also progressed greatly since the proposition that a conserved RGD motif in the FMDV loop binds the virus to cells via an integrin (Fox et al. 1989; Pfaff et al. 1988; Surovoi et al. 1988). Two classes of receptors are now recognised: integrins and heparan sulphate proteoglycans (Berinstein et al. 1995; Jackson et al. 2000b, 1996, 2002; Neff et al. 2000). Interactions with the latter have been detailed via structural studies of two virus-heparan sul-

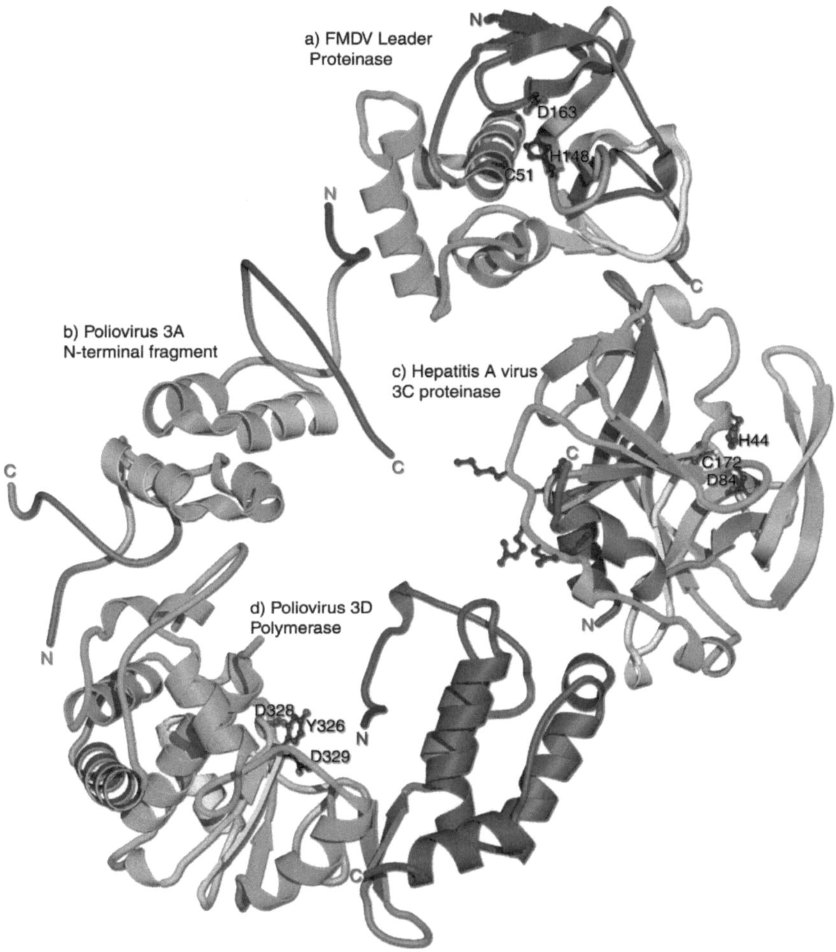

Fig. 4 Non-structural proteins. **a** The FMDV leader proteinase (Guarné et al. 2000) (a papain-like cysteine proteinase) is drawn as a *ribbon*, colour-ramped from *blue* to *red* from the N- to C-terminus. The secondary structure elements were defined with RASMOL. The catalytic residues [His148 and Cys51 (mutated to an Ala) and the proposed third member of the catalytic triad (Asp163)] are depicted in *ball-and-stick* with standard atom colouring. **b** The dimeric N-terminal fragment of polio 3A (Strauss et al. 2003) drawn as **a**. **c** Hepatitis A virus 3C proteinase (Bergmann et al. 1997) drawn as **a**. The nucleophile Cys A172 and the acid-base catalyst His A44 are depicted in *ball-and-stick* with standard atom colouring together with Asp84 (originally thought to be the third member of the catalytic triad). Residues 95–99 KFRDI (the RNA recognition sequence) are depicted in *blue ball-and-stick*. **d** The poliovirus 3D polymerase (Hansen et al. 1997) drawn as **a**. The conserved metal binding motif 4xDD (residues 326–329) is depicted in *ball-and-stick* with standard atom colouring

phate receptor complexes (Fry et al. 1999) (Fig. 2). The molecular processes that follow receptor recognition remain unclear, but the capsid structure of FMDV has provided an explanation for the observation that FMDVs share with the cardioviruses a mechanism of dissociation distinct from rhino- and enteroviruses.

The capsid proteins constitute approximately one-third of the proteins encoded by the virus genome. Of the remaining proteins which regulate viral replication, protein processing and modifications of the host, structural information is available for the leader proteinase of FMDV (Guarné et al. 2000), the 3C proteinases of related picornaviruses (Matthews et al. 1994; Allaire et al. 1994; Mosimann et al. 1997), the 3D polymerase of poliovirus (Hansen et al. 1997), the soluble domain of 3A (Strauss et al. 2003) of poliovirus and the 2A proteinase of rhinovirus (Petersen et al. 1999) (Fig. 4). Knowledge from all of these structures will be drawn upon, bearing in mind that, although many aspects of protein replication will be similar for related picornaviruses, studies of the capsid proteins and comparison of the genomes reveal significant differences between the genera.

2 Structural Overview

In the mature virus particle (Fig. 1), 60 copies of each of the four structural proteins VP1–4 associate to form the icosahedral shell or capsid. The virion has a pseudo T=3 icosahedral architecture, made possible by the broadly similar structures of VP1, VP2 and VP3. The arrangement is similar to that observed in other picornaviruses where the VP1 proteins are located around the icosahedral five-fold axes, VP2 and 3 alternate around the three-fold axes and the shorter VP4 is located entirely at the inner surface of the capsid. VP1–3 adopt a fold almost ubiquitous in RNA viruses [hence often referred to as the RNA virus fold (RVF)], that of a wedge-shaped eight-stranded β-barrel. According to convention the strands of the two sheets are labeled alphabetically proceeding from the N- to the C-termini of the amino-acid sequence, namely CHEF and BIDG. These β-wedges form a well-fitting mosaic that comprises the bulk of the capsid. Loops connecting the strands form the exterior surfaces and are identified by the strands they join; thus the hypervariable sequence known as the FMDV loop, spanning residues 140–160 of VP1, is also known as the GH loop (Fig. 2). For FMDV, the major structural proteins, VP1–3 are smaller than their counterparts in other picornaviruses, especially so VP1, each having a molecular weight of approximately 24,000 (Fig. 3). This reduction in size is achieved by the truncation of surface loops rendering the capsid both thinner (excluding VP4 the average capsid thickness is 33 Å as compared to 42.5 Å for human rhinovirus 14) and smoother, with none of the remarkable surface features such as the pits and canyons described for other picornaviruses (Hogle et al. 1985; Rossmann et al. 1985; Luo et al. 1987). In fact, for FMDV the C-terminus of VP1 traverses the biological protomer in a clockwise direction finishing adjacent to the VP1 GH loop of the 5-fold related protomer and in doing so fills the depression which would be analogous to the canyon/pits.

FMDV serotypes have on average 86% sequence identity to each other, although VP1 is substantially more variable. Whereas most structural differences between subtypes are confined to side chains and resemble the changes observed in MAR mutants (Parry et al. 1990; Lea et al. 1995), a comparison of serotypes (Acharya et al. 1989; Lea et al. 1994, 1995; Curry et al., 1996) reveals more significant main-chain differences, providing some structural basis for serotype and subtype differentiation (Fig. 3).

3
Capsid Stability and Permeability

Of the major capsid proteins, VP1 contributes most to the accessible surface of the virus whereas VP3 contributes most towards capsid stability (Fig. 2). The N-termini of VP3 associate to form a β-annulus at the icosahedral 5-fold axes, knitting the protomeric subunits (VP1–4) into pentamers and leaving a small pore or channel along the 5-fold axes which is accessible from the capsid exterior. In common with nearly all other picornaviruses, FMDV possesses a myristoyl group (Chow et al. 1987; Belsham et al. 1991) covalently linked to the N-terminus of VP4. This is not visible in the X-ray structures of FMDVs, but, by analogy with other picornavirus structures (Hendry et al. 1999), it is likely to form a cluster at the base of the icosahedral 5-fold axes, possibly stabilizing the N-terminal strand of the VP3 β-barrel. Further stability is provided by a feature unique to FMDV, a ring of disulphide bonds around the 5-fold axes towards the virus exterior, linking the VP3 N-termini and constricting the channel. VP1-VP2 covalently associated dimers are also seen in serotype O viruses, where they have a special role in modulating the mobility of the VP1 GH loop (Logan et al. 1993). The capsid as a whole is fairly permeable; FMDV particles have the highest buoyant density of any picornavirus in CsCl and permit the entry of proflavin, the dimensions of which are commensurate with its entry via the 5-fold channel (Acharya et al. 1989).

Capsid integrity is vulnerable to low pH; below pH 6.8, the capsids dissociate to 12S pentameric assemblies (Brown and Cartwright 1961) and this property probably influences virus targeting, pathogenicity and spread (Fig. 1). It has been proposed that the pH instability of FMDV is governed by the switch of the protonation state of a cluster of histidines within 15 Å of the pentamer boundary (van Vlijmen et al. 1998), in particular the interactions of His142 and His145 of VP3 with polar residues across the inter-pentamer boundary. His142 was proposed (Curry et al. 1995) to mediate a histidine-α-helix charge-dipole interaction, a role verified by mutational analysis (Ellard et al. 1999), although electrostatic calculations suggest that this interaction contributes only a small proportion of the destabilisation energy. An extended β-sheet of six strands straddles the pentamer boundaries, composed of four strands from VP3 (pentamer 1) and two (β-A1 and β-A2) from VP2 (pentamer 2). Together with cation ligation on the three-fold axis this maintains capsid integrity. To date, no structural information is available for dissociated pentamers. In rhino- and enteroviruses; the trans-pentamer β-sheet structure is strengthened by a seventh strand contributed by the N-terminal extension of VP1 of pentamer 1 forming a sandwich around the strands contributed by pentamer 2 (Acharya et al. 1990; Fry et al. 1991, 1990). The rhino- and enteroviruses do not readily dissociate into pentameric subunits but rather uncoat via A-particle intermediates (Rueckert 1996) in which VP4 and the N-terminus of VP1 are externalized. A final major, and biologically important, structural distinction divides the FMDVs and cardioviruses from rhino- and enteroviruses. In the latter viruses the β-barrel of VP1 contains a long, hydrophobic pocket that accommodates a fatty acid molecule, often referred to as the pocket factor (Filman et al. 1989; Lewis et al. 1998). The release of this molecule on receptor binding triggers viral uncoating in the rhino- and enteroviruses, whereas FMDVs and cardioviruses dissociate via a different mechanism and have no trace of, nor room for, a pocket factor (Fry et al. 1991).

4
Genome Penetration

A common problem for all picornaviruses is that of transposing their RNA genomes across the cell membrane to initiate the infection process in the cytoplasm. For poliovirus and rhinoviruses conformational changes in the virus capsids, which are induced by receptor binding and/or internalisation into endosomes, result in the loss of the RNA and VP4 without disruption of the particles. There is increasing evidence that these conformational changes result in the insertion of VP4 and the N-terminus of VP1 into the cellular membranes to produce channels through which the genomic RNA might traverse the membrane (Hogle 2002). The infection process by FMDV appears to be markedly different, and there is no model at present as to how the genome enters the cell. All of the available evidence suggests that the process is triggered simply by the acidic environment of the endosome. Acid treatment results in the dissociation of the virus into pentameric subunits, VP4 and the genome, and available evidence suggests that the capsid proteins remain in endosomes while the RNA penetrates into the cytoplasm. Acidification of the virus in vitro produces an insoluble aggregate of VP4 while the pentmeric 12S subunits remain soluble. It is possible, therefore, that VP4 may associate with the endosomal membrane to produce channels into the cytoplasm, but at present this is purely speculative.

5
Immunogenicity/Receptor Binding

Multiple antigenic sites have been identified for the majority of picornaviruses studied (Boege et al. 1991; Hogle and Filman 1989; Sherry et al. 1986). In general, the entire accessible surface of a virus would be expected to be antigenic, although antigenic variation is likely to be more apparent in residues which are not essential for capsid structure or function. For type O1 FMDVs, up to five antigenic sites have been described (Barnett et al. 1989; Crowther et al. 1993; Kitson et al. 1990) involving all three major capsid proteins (Fig. 2). For C serotype viruses, the identification of sites is similar, with a major discontinuous antigenic site located near to the three-fold axis involving residues of VP1, VP2 and VP3 (Lea et al. 1994). Two major antigenic sites have been located on VP1 and VP3 of A10 (Thomas et al. 1988) and corresponding sites identified on A12 (Baxt et al. 1989). There is an underlying structural similarity between the antigenic sites of the different serotypes, revealed by early studies which demonstrated the key importance of mobile, exposed regions on a single FMDV capsid protein, VP1, for the biological properties of the virus. In all FMDVs the C-terminus and the GH loop of VP1 are highly exposed regions which are central to both antigenicity and receptor binding (Barnett et al. 1989; Baxt and Becker 1990; Crowther et al. 1993; Fox et al. 1989; Jackson et al. 1997, 2000a, 2000b, 2002; Kitson et al. 1990; Liebermann et al. 1991; Mason et al. 1994; McCahon et al. 1989; Parry et al. 1989, 1985; Stave et al. 1988; Surovoi et al. 1988; Xie et al. 1987). In type O FMDV, for example, trypsin cleavage of VP1 within the GH loop and at the C-terminus does not affect the integrity of the virus particle but abolishes the ability of the virus to attach to cells and greatly reduces immunogenicity.

The VP1 GH loop has been found to be structurally disordered in all serotypes studied to date (Acharya et al. 1989; Curry et al. 1996; Lea et al. 1994) (Fig. 3). The mobility of this loop presumably presents more possibilities for interaction with antibodies, leading to its extreme antigenicity. Nevertheless, for certain variant viruses which escape neutralisation by a monoclonal antibody which bound simultaneously to the GH loop and to the C-terminus of VP1, escape substitutions were, surprisingly, found in a quite different part of the virus, the VP1 BC loop (Parry et al. 1990). Crystallographic studies of these mutants and comparisons with the parent virus revealed that single amino acid changes in the VP1 BC loop were responsible for switching the conformation of the VP1 GH loop. The electron density profiles suggested that the VP1 GH loop exists in two predominant conformations, "down" and "up", and that the conformational restriction imposed by the disulphide tethering one of the ends of the loop in O1 BFS (C134 of VP1-C130 of VP2) was responsible for driving it towards the less intrinsically stable "up" conformation in the parent virus. Thus the substitutions in the VP1 BC loop underlying this "up" conformation were able to destroy the integrity of the VP1 GH loop epitope by switching it to a predominantly "down" conformation. On the basis of these data, the VP1 GH loop was proposed to act as a hinged domain (Parry et al. 1990). This hypothesis was confirmed when the VP1 GH loop structure was visualised in a well-ordered "down" conformation by the crystallographic analysis of DTT-reduced O1BFS virus (Logan et al. 1993), in which the destabilizing Cys134 VP1–Cys130 VP2 disulfide was deliberately broken. In the reduced virus structure, the VP1 GH loop lies predominantly over the surface of VP2 (the C_α for Cys134 of VP1 moves some 12 Å). The N-terminal portion of this loop (residues 130–160) traverses the biological protomer in a clockwise direction, coming into close proximity with the VP3 GH loop (residues 173–180) from a five-fold related biological protomer, resulting in a rearrangement of that loop. Thus observed conformational differences in this VP3 GH loop between serotypes are more likely to be attributable to the flexibility of the VP1 GH loop than to genuine serotype specific changes (Curry et al. 1996). The VP1 GH loop then forms a strand of β-sheet adjacent to strand C of VP2. The RGD motif (residues 145–147) occupies a turn prior to a 3_{10} helix and is in an extended conformation (Fig. 2), similar to that observed for the same motif in γ-crystallin, which is known to bind integrin (Wistow et al. 1983).

Although in native viruses the VP1 GH loop is disordered, there is evidence that the internal structure seen in reduced viruses is preserved in different loop positions. Thus the crystal structure of a synthetic peptide corresponding to the sequence of the loop for a serotype C virus in complex with an Fab fragment from a neutralising monoclonal antibody revealed a loop structure similar to that seen in crystals of reduced O1BFS (Verdaguer et al. 1995). Each of these loop structures was also found to be maintained after multiple amino acid substitutions (Lea et al. 1995; Ochoa et al. 2000). Such conservation suggests that the conformation as well as the chemical properties of these residues is important in integrin recognition.

Because FMDV replicates in the reducing environment of the cytoplasm, O1 strains are initially produced in the reduced state, and so presumably possess a VP1 GH loop that is ordered. Oxidation to the mature, disordered form occurs only after release from the host cell. Indeed, studies of the re-oxidation of reduced virus showed that the disulphide bond reforms. The ability to form the disulphide bond despite the large deviation in the position of residue 134 of VP1 emphasises the residual mobility of the reduced loop as evidenced by the high crystallographic B-factors (measures of thermal motion) and the estimated 85% occupancy compared to the rest of the structure. Not surprisingly, the infectivity of reduced virus with an 'ordered loop' is markedly lowered, unless the virus can use an alternative receptor (A. King, unpublished data). The release of virus initially unable to bind receptor might be important for pathogenesis, facilitating viral spread.

Whereas the reduced virus structure confirmed the "down" position for the loop proposed by Parry et al.(1990), an Fab-C-serotype virus complex (Hewat et al. 1997) determined by cryoelectron microscopy provided direct evidence for the "up" position, because this accounted best for the observed position of the bound Fab (Fig. 2). Analysis of a further C serotype virus-Fab complex and the corresponding Fab-peptide complex (Verdaguer et al. 1999) showed the VP1 GH loop in a similar conformation but disposed differently on the virus surface, a difference explicable by a simple rotation hinged about the base of the loop (Hewat et al. 1997; Verdaguer et al. 1999). Together these results verify the hypothesis that the loop exists as a self-contained, biologically important folding unit which can occupy at least two radically different positions on the virus surface (Acharya et al. 1989; Parry et al. 1990) (Fig. 2). The combination of flexibility and rigidity in the structure of the VP1 GH loop for native viruses facilitates receptor binding via the conserved RGD integrin-binding motif within it.

The strong antigenicity of this receptor binding site is in complete contrast to what would be expected from the canyon hypothesis for receptor binding proposed for other picornaviruses (Luo et al. 1987; Rossmann et al. 1985), in which the residues involved in cell attachment are located at the base of surface depressions whose dimensions render them accessible to cell receptors but not antibodies. In contrast, the mode of neutralisation for anti-FMDV antibodies recognising this site is clearly by direct blocking of receptor attachment (Verdaguer et al. 1995) and all the antigenic sites flank the RGD triplet [if we consider both of the predominant conformations (Fig. 2)]. Taking into account the large footprint of an antibody (typically 700 $Å^2$), we would expect that variation around the conserved attachment site would be sufficient to evade immune surveillance. In fact, more subtle effects also operate, as we have seen, such that a single mutation in the structure underlying the VP1 GH loop (e.g. residue 82 of VP2 or residue 43 or 59 of VP1) can perturb its structure, resulting in an amplified propagation of structural change (Curry et al. 1996; Parry et al. 1990).

Because the receptor attachment site is at a peripheral 'flexible' location, receptor binding is unlikely to mediate any capsid dissociation by imposing mechanical strain on the capsid [as is thought to occur in other picornaviruses (Rueckert 1996)], hence the requirement for an additional factor, acid lability, to promote disassembly (Acharya et al. 1989).

6
Receptor Interactions

FMDV targets epithelial cells, and much progress has been made in identifying cellular receptors for FMDV (Baranowski et al. 2000; Berinstein et al. 1995; Jackson et al. 1996, 2000b, 2002; Neff et al. 2000).Two classes of receptors have been recognised, integrins and heparan sulphate proteoglycans. The most favoured integrin isoforms are $\alpha v\beta 3$, $\alpha v\beta 6$ and $\alpha v\beta 1$ (Jackson et al. 1997, 2000b, 2002). It has also been shown that antibody-bound FMDV can infect cells via Fc receptor-mediated adsorption (Mason et al. 1993) and via an engineered artificial receptor consisting of a single-chain anti-FMDV monoclonal antibody fused to ICAM-1 even with RGD-deleted viruses, suggesting that regions of the virion other than the G-H loop of VP1 might bind to and infect cells in culture.

7 Heparan Sulphate Proteoglycan Receptors

Heparan sulphates (HSs) are polymers of disaccharide repeats of L-iduronic acid and D-glucosamine, which carry random patterns of sulfation and are hence negatively charged (Bernfield et al. 1999; Kjellen and Lindahl 1991). They are found as the carbohydrate component of many proteins called heparan sulphate proteoglycans (HSPGs). They are expressed on the surface of virtually all cell types, either as integral membrane proteins or as components of the extracellular matrix.

HS was originally identified as a potential enhancer (or co-receptor) of cell entry by certain strains of type O FMDV (Jackson et al. 1996). Subsequently, viruses representing several other serotypes, including serotypes A (Fry et al. 1999), C (Baranowski et al. 1998, 2000; Escarmis et al. 1998), Asia-1 and SAT-1 (T. Jackson, unpublished data) have been shown to bind HS and/or use HSPGs as cellular receptors. It was at one time thought that HS acts as the virus' first point of contact with the cell, en route to its integrin receptor (Jackson et al. 1996). However, a functional relationship between HS and integrins was not established, and it is now clear that propagation of certain strains of FMDV in cultured cells selects for virus variants that have a high affinity for HS (Sa-Carvalho et al. 1997). Such viruses can then completely dispense with their RGD integrin-binding site and use HSPGs as alternative receptors without the mediation of integrins (Baranowski et al. 2000; Martinez et al. 1997; Neff et al. 1998).

Viruses that bind to HS arise rapidly in cell culture and are characterised by having a small plaque phenotype and an increased virulence and expanded host range for cultured cells typified by an acquired ability to infect CHO and K562 cells (Baranowski et al. 2000; Jackson et al. 1996; Neff et al. 1998; Sa-Carvalho et al. 1997). Remarkably, the ability of

FMDV to bind HS arises from only one or two residue changes on the outer capsid surface which result in a net gain in positive charge (Fry et al. 1999; Sa-Carvalho et al. 1997). Subtype O strains, for example, have a histidine residue at VP3-56 which changes to arginine on passage in tissue culture, and this single change was identified as important for HS binding by genetic studies using a cloned infectious copy of the genome (Sa-Carvalho et al. 1997). For a single substitution to confer such a dramatic change in HS affinity suggests that the icosahedral virion, with 60 binding sites, interacts with HS polymers at multiple sites, so amplifying the effect of any sequence changes, although direct crystallographic evidence demonstrates that monovalent interactions can be significant (see below).

The question arises as to whether the ability of FMDV to switch rapidly between its integrin and HS receptors plays any role in FMDV pathogenesis. Despite conferring a clear advantage for growth in cultured cells, viruses that have a high affinity for HS are attenuated for cattle (Sa-Carvalho et al. 1997). Furthermore, infection of cattle by an HS-binding strain gave rise to virulent revertant viruses which had lost the ability to bind HS. These observations imply that a high affinity for HS is a disadvantage for FMDV in an animal. It remains to be established whether low-affinity interactions between FMDV and HS occur in vivo, or whether the ability of FMDV to switch rapidly from integrin to HS receptors plays any role in the latter stages of FMDV pathogenesis, once an infection has been established (Fry et al. 1999).

8 The Structural Basis of Heparan Sulphate Recognition

The crystal structures of two tissue culture-adapted strains of FMDV, O1BFS (Fry et al. 1999) and A10 (E.E. Fry, unpublished data), complexed with HS have been determined. The HS binding site in both complexes is formed by a shallow depression in the centre of the biological protomer similar to the putative receptor-binding 'pit' of cardioviruses (Luo et al. 1987). The models reveal a segment up to five sugar residues binding in a position, orientation and conformation which are almost identical for the two serotypes (Fig. 2). The bound sugar motif is fully sulphated and makes contact with approximately nine amino acid residues derived from all three major structural proteins (VP1–VP3). A comparison with the unliganded virus structures shows that virtually no conformational changes occur in either virus on HS binding. Arg56 of

VP3 occupies a critical position in both complexes, making ionic interactions with two sulphate groups. This residue switches to Arg from the wild-type His when O1 viruses adapt to tissue culture (Sa-Carvalho et al., 1997). A second contact residue, Arg 135 of VP2, which is also conserved between O1BFS and A10, plays a subsidiary role in interacting via water molecules with one of the HS disaccharides. Most other contact residues are not conserved between these two viruses (E.E. Fry, unpublished data). In contrast, studies with type C FMDV (Baranowski et al. 2000; Escarmis et al. 1998) have identified residues involved in HS adaption at widely spaced locations on the capsid, leading the authors to suggest that there may be more than one potential binding site in this serotype.

The C-terminal residues of VP1 (residues 201–211) have also been implicated in cell attachment because selective removal of these residues by treatment with the lysine-specific endoproteinase Lys-C results in virus particles which are no longer capable of binding to cells (Fox et al. 1989). It is unclear whether this region is involved in binding to HS or to integrin receptors. The sequence of this region is similar to the heparin-binding site of vitronectin, leading to the suggestion that there is a direct interaction with HS. However, the crystal structure of the virus-HS complex demonstrated that this is not the case (Fry et al. 1999). Nevertheless, the nearby residues, His 195 and Lys 193 of VP1 of O1BFS and A10, respectively, do make contact with HS and it is possible that the C-terminus of VP1 may serve to stabilise these residues in a suitable position for HS binding. Interestingly, the HS binding site is also one of the five antigenic sites (site 4; Fig. 2). As the antibody that mapped to this site was raised against the wild-type (i.e. integrin dependent) form of the virus, its mechanism of neutralisation may not be via receptor blocking.

The heparan sulphate receptor binding site is located some 15 Å away from the RGD motif in the reduced conformation. The adaptation to HSPG receptors by O1 viruses appears not to compromise their ability to make use of integrin receptors, implying that the two attachment sites function independently of each other (S. Najjam, unpublished data).

Overall, we feel that the similarities in the crystal structures of the O and A type virus-HS complexes suggest that this binding site has an important role in the survival of the virus, perhaps via persistent infections (Fry et al. 1999).

9
Integrin Receptors

Integrins are heterodimeric proteins composed of two, differing, type 1 transmembrane subunits, α and β, with large extracellular domains and usually a short cytoplasmic tail and are the key determinants of cell entry by FMDV. Most integrins recognise their ligands by binding to short linear peptide sequences, and several members of the integrin family, including $\alpha v\beta 1$, $\alpha v\beta 3$, $\alpha v\beta 5$, $\alpha v\beta 6$ and $\alpha 5\beta 1$, recognize their ligands by binding to the tripeptide RGD. Of these $\alpha v\beta 5$ and $\alpha 5\beta 1$ have not been shown to act as receptors for FMDV (Baranowski et al. 2000; Jackson et al. 2000b; Mason et al. 1993; Neff et al. 1998), and according to the distribution of the others in tissues, $\alpha v\beta 6$ is most likely to correspond to the true cellular receptor (though the distribution of $\alpha v\beta 6$ in bovine tissues remains unclear). FMDV is thought to enter the cell through endosomes where capsid uncoating is triggered by the acidic environment of this compartment (Curry et al. 1995; Ellard et al. 1999; Mason et al. 1993) and infection is inhibited by pre-treatment of cells with reagents that raise endosomal pH (Baxt 1987; Carillio et al. 1984; Miller et al. 2001). Mutant $\alpha v\beta 6$ receptors with deletions in a conserved motif (known to function in other internalisation receptors as a signal which directs membrane proteins into clathrin-coated vesicles) (Chen et al. 1990; Trowbridge et al. 1993) are defective at mediating infection despite retaining the ability to bind FMDV (Miller et al. 2001), suggesting that the $\beta 6$ cytoplasmic domain may contain signals necessary for virus uptake into endosomes.

Several lines of evidence show that the residues following the RGD motif of FMDV, including those at the RGD +1 and +4 positions, are important for receptor recognition (Rieder et al. 1994, Mateau 1996, Jackson et al. 2000a). The similarity between the RGD site of FMDV and LAP-1 and the observation that the DLXXL motif is a ligand for $\alpha v\beta 6$ (Kraft et al. 1999) suggest that conservation of the leucine residues at the RGD +1 and RGD +4 positions is likely to reflect the use of $\alpha v\beta 6$ rather than $\alpha v\beta 3$ as a specific receptor. A model of the virus with the loop in the "up" position can readily be docked with the extended $\alpha v\beta 3$ (Xiong et al. 2001) structure in a fashion analogous to the Fab interaction.

10 Virus-Integrin Complexes

To date no complex of FMDV with integrin has been visualised. Perhaps the closest approximation comes from work with hepatitis B cores expressing the FMDV loop, which have been visualised decorated with $\alpha v\beta 3$ by EM (Sharma et al. 1997).

Nevertheless, the recent report of a crystal structure for the ectodomain of $\alpha v\beta 3$ should, in the future, facilitate interpretation of cryo-EM reconstructions if these can be obtained.

11 Non-structural Proteins

For FMDVs, the product of translation of the single-stranded positive-sense RNA genome is a polyprotein that is subsequently processed by a variety of virus-encoded 'proteases' (Leader, 2A, 3C).

The leader proteinase L^{pro} (201 residues in serotype O) is found at the N-terminus of the polyprotein in aphthoviruses [closely conserved between the distantly related equine rhinovirus A (ERVA) and FMDV (Hinton et al. 2002)]. It is synthesised in two forms, Lab and Lb, the latter predominating in vivo. A papain-like proteinase, it mediates autocatalytic cleavage of itself from P1 and cleaves the host protein eIF4GI, resulting in the shut-off of host cap-dependent mRNA translation. A similar function is performed by protein 2A in human rhino- and enteroviruses, although the cleavage occurs at a different site. L^{pro} appears to play a critical role in pathogenesis. An X-ray crystallographic study of L^{pro} (Guarne et al. 2000) (Fig. 4) confirms that the active site triad comprises Cys51, His148 and Asp163 (the latter in place of the Asn usually found in papain-like proteinases). In addition, it is unusual in that the His-Asp interaction is exposed to solvent and that the transition state analogue is stabilized by an Asp (Guarne et al. 2000). Other notable differences to papain include a high sensitivity to increases in cation concentration and activity limited to a narrow pH range. The structure suggests that these characteristics are due to the presence of Asp49, 163 and 164 close to the active site (not normally present in papain-like enzymes).

There is as yet no structural information for proteins and their intermediates derived from the P2 and P3 regions of the FMDV genome, but some information is available for analogous P3 proteins from rhinovirus and poliovirus and these may provide a paradigm for picornaviruses. These proteins participate in RNA replication and protein folding and assembly.

In aphtho- and cardioviruses, the primary polyprotein cleavage occurs not between P1 and 2A but at the C-terminus of 2A. The three C-terminal residues of aphtho- and cardiovirus 2A are completely conserved (-NPG-), as is the N-terminal residue (proline) of 2B (Palmenberg et al. 1992). Although FMDV 2A comprises only 18 amino acids, it appears to function as an autoproteinase [as do the 19 C-terminal residues of cardiovirus 2A (Donnelly et al. 1997)]. However, it has been hypothesized for FMDV 2A that it actually acts at the level of the ribosome, using a novel mechanism to modify the translational machinery, allowing the release of 2A whilst permitting the synthesis of the downstream proteins (Donnelly et al. 2001). Thus the structure of the rhinovirus 2A protease (Petersen et al. 1999) is probably irrelevant to FMDV 2A.

Picornaviral 2B (154 residues type O) and 2C (318 residues type O) proteins localise to the ER–derived outer surface vesicles which are the sites of genome replication (Lubroth and Brown, 1995). In poliovirus, 2B appears to enhance membrane permeability and block secretory pathways whereas 2C is required for the initiation of negative-strand RNA synthesis. In all picornaviruses, these proteins contain conserved nucleoside triphosphate-binding and helicase motifs, have ATPase and GTPase activity and are highly conserved.

For the P3 proteins 3A and 3B are anomalous in FMDV. 3A is almost twice as long in FMDV (143 residues) as in poliovirus. A C-terminal truncation produces a bovine-attenuated phenotype which is highly virulent in pigs (Beard and Mason 2000) and adaptation to guinea-pigs is conferred by a single amino acid change in 3A, implicating it in host adaptation. FMDV 3A has been shown to co-localize with the intracellular membrane system where replication occurs (O'Donnell et al. 2001). This correlates with the prediction of a transmembrane helix between residues 59 and 76. 3A also forms stable intermediates 3AB and 3ABB. For poliovirus, 3AB can bind the 5′ cloverleaf structure of the viral genome (Xiang et al. 1995) and can directly bind to and act as a co-factor of 3D (Richards and Ehrenfeld 1998, Xiang et al. 1998). An NMR structure for the soluble N-terminal fragment of poliovirus 3A (Fig. 4) is available (residues 1–59 of 87) (Strauss et al. 2003). This structure consists of a symmetrical amphipathic α-helical hairpin dimer with unstructured termini and a charged surface patch, presumably the site of RNA interaction. The hydrophobic portion may contribute to a second dimeric interface in line with pore formation and its ability to permeabilise membranes.

3B (VPg) is covalently bound to the 5′ terminus of the genome and implicated in priming replication. The first step in poliovirus RNA synthesis is the covalent linkage of UMP to 3B (Paul et al. 2003). FMDV unusually possesses three non-identical copies of 3B in tandem (23–24 amino acids apiece) with a highly conserved Tyr at position 3 which mediates the phosphodiester linkage to the viral RNA (Forss and Schaller 1982). Not all three copies are required for replication (Falk et al. 1992) but they are clearly important to the virus, possibly playing a role in pathogenesis or host-range (Pacheco et al. 2003).

FMDV $3C^{pro}$ (213 residues) can efficiently process all 10 cleavage sites in the polyprotein (Bablanian and Grubman 1993). Whereas in poliovirus, $3CD^{pro}$ has been implicated as the major viral proteinase, responsible for cleavage of the structural protein precursor, in FMDVs levels of 3CD are variable and it appears to be rapidly cleaved. FMDV $3C^{pro}$ shows greater heterogeneity in cleavage sites than poliovirus $3C^{pro}$. It can also cleave the host proteins eIF4A (which is part of the cap-binding complex and functions as an RNA helicase) and eIF4G, late in the infectious cycle and at a different site to L^{pro} (Belsham et al. 2000), and it may inhibit host cell transcription by cleavage of histone H3.

The sequences of $3C^{pro}$s show that they belong to the chymotrypsin-like family of serine proteases, and the high degree of conservation suggests that the three-dimensional structures for rhinovirus, hepatitis A and poliovirus (Matthews et al. 1994; Allaire et al. 1994; Mosimann et al. 1997) should provide reasonable models for other members of the family (Fig. 4). However, there is a difference in the nature of the catalytic site whereby the family divides into two lineages. The entero- and rhinoviruses have a Glu as a member of the serine proteinase-like catalytic triad which interacts with a His, as observed in the HRV14 structure. The Asp which replaces the Glu (conserved in lineage B) was observed in the hepatitis A 3C proteinase structure not to interact with the histidine (Allaire et al. 1994), a water molecule fulfilling this role. Biochemical studies have identified the residues involved in the catalytic triad in FMDV 3Cpro as Cys163, His46 and Asp84 (Grubman 1995). Thus FMDV is a member of the second lineage and we would expect the active site to more closely resemble the hepatitis A 3C proteinase structure (Fig. 4).

$3CD^{pro}$ has been shown to bind to positive-strand RNA (Andino et al. 1990, 1993). Mutations shown to affect RNA-binding have been located on the three-dimensional structure of $3C^{pro}$, revealing the binding site to be on the opposite side to the catalytic site and in close proximity to the N- and C-termini such that RNA binding may regulate protein processing.

FMDV 3D (470 residues) is an RNA-dependent RNA polymerase which, together with other viral proteins and possibly host proteins, carries out viral RNA replication to generate new template, messenger and viral RNAs from the original infecting RNA. This appears to take place within a membranous complex. There is a high degree of conservation among the different FMDV sero- and subtypes for 3D at both the nucleotide and amino acid sequence levels. The structure of poliovirus 3D (Hansen et al. 1997) (Fig. 4) revealed a strong similarity to other classes of polymerases with the characteristic 'palm', 'fingers' and 'thumb' subdomains. The core 'palm' subdomain comprises four of the conserved sequence motifs found in other polymerases. The active site consists of the conserved YGDD sequence and a highly conserved D residue also found in this subdomain. Interestingly, the crystal structure also revealed polymerase-polymerase interactions at two interfaces which appear to be functionally significant (Hobson et al. 2001). This tallies with the high degree of cooperativity in RNA binding and polymerisation activities, suggesting that the polymerase may function as an oligomer (Beckman and Kirkegaard 1998). This could produce multiple interactions with other viral and host cell factors consistent with a recent model of initiation of poliovirus negative-strand RNA synthesis which includes a poly (A)-binding protein (PABP)-3CD complex and a PCBP-3CD complex that interact with each other to form a circular RNP complex (Barton et al. 2001, Herold and Andino 2001).

12 Capsid Assembly

Structural proteins accounting for approximately one-third of the polyprotein are encoded towards the 5′ end of the open reading frame (the P1 region) and appear to fold into a structure which is antigenically very similar to the virus even prior to the cleavages which yield the viral proteins VP0, VP1 and VP3 (also known as 1AB, 1D and 1C, respectively).

The P1/2A cleavage to form the structural protein precursor, P1 (Fig. 1), is performed by the 3C protease. P1 folds into a protomer, five of which associate to form a pentamer after the VP2/3 and VP3/1 cleavages have taken place. We have no structural information for either of these intermediates. Beyond this point the assembly and encapsidation mechanism is unclear; either the pentamers assemble into empty capsids and the RNA [covalently linked to VPg (3B)] is inserted to form the provirion (the immature particle in which the maturation cleavage of VP0 has yet to occur) or the pentamers interact directly with the RNA/VPg, forming the provirion without an empty capsid intermediate. More recent studies with poliovirus (Verlinden et al. 2000) favour the latter hypothesis whereby empty capsids are either a failed product or serve as a repository of pentamers. The final step in producing the mature virion is the cleavage of 1AB (VP0), which occurs via a presumed autocatalytic mechanism in the presence of viral RNA, converting the N-terminal 85 residues into VP4 and the remaining residues into VP2 (1A and 1B, respectively), although a few copies of 1AB may remain in the intact virion.

Structural analyses of empty capsids of type A22 FMDV (Curry et al. 1997) (which despite lacking RNA possess cleaved VP0) reveal that increased ordering of the N-terminus of VP1 and C-terminus of VP4 in the vicinity of the 3-fold axes is associated with RNA encapsidation and that a further stabilization of the capsid is attributable to VP0 cleavage. A conserved histidine (His145 of VP2) could mediate the autocatalytic VP0 cleavage in FMDV (Curry et al. 1997), as predicted in poliovirus for the analogous His195 of VP2. Naturally occurring FMDV provirions have not been observed; however, noninfectious mutated provirions with uncleaved 1AB have been generated by reverse genetics and these remain acid sensitive.

13 Concluding Remarks

The small size, icosahedral geometry and relative simplicity of FMDV lend it to detailed structural analysis. Thus far, these factors have enabled much progress in understanding in intimate detail how the virus binds its cellular receptors and uncoats during the process of cell entry, and also how neutralising antibodies recognise the particle. It now remains to obtain a clearer picture of the processes involved in virus replication in atomic detail. FMDV was the first animal pathogen to be identified as a virus, and today, more than a century later, it remains at the forefront of structural virology. Future progress towards more humane and cost-effective ways of controlling FMD will benefit from a better understanding of the molecular mechanisms underlying infection and transmission of this most infectious of all viruses.

References

Acharya, R., Fry, E., Logan, D., Stuart, D., Brown, F. and Rowlands, D. (1990) New aspects of positive strand RNA viruses. In: M.A. Brinton and F.X. Heinz (Eds), Some observations on the three-dimensional structure of foot-and-mouth disease virus

Acharya, R., Fry, E., Stuart, D., Fox, G., Rowlands, D. and Brown, F. (1989) The three-dimensional structure of foot-and-mouth disease virus at 2.9-Å resolution. Nature (London) 337, 709–716

Adams, P., Lea, S., Newman, J., Blakemore, W., Najjam, S., King, A., Stuart, D. and Fry, E. A three-dimensional structure for foot-and-mouth disease virus SAT 1 serotype. (in preparation)

Allaire, M., Chernaia, M.M., Malcolm, B.A. and James, M.N. (1994) Picornaviral 3C cysteine proteinases have a fold similar to chymotrypsin-like serine proteinases. Nature 369, 72–6

Andino, R., Rieckhof, G.E., Achacoso, P.L. and Baltimore, D. (1993) Poliovirus RNA synthesis utilizes an RNP complex formed around the 5′-end of viral RNA. EMBO J. 12, 3587–98

Andino, R., Rieckhof, G.E. and Baltimore, D. (1990) A functional ribonucleoprotein complex forms around the 5′ end of poliovirus RNA. Cell 63, 369–80

Bablanian, G.M. and Grubman, M.J. (1993) Characterization of the foot-and-mouth disease virus 3C protease expressed in *Escherichia coli*. Virology 197, 320–7

Baranowski, E., Ruiz-Jarabo. C. M., Sevilla. N., Andreu. D., Beck, E. and E. Domingo, E. (2000) Cell recognition by foot-and-mouth disease virus that lacks the RGD integrin-binding motif: Flexibility in Aphthovirus receptor usage. J. Virol. 74, 1641–1647

Baranowski, E., Sevilla, N., Verdaguer, N., Ruiz-Jarabo, C. M., Beck, E. and Domingo, E. (1998) Multiple virulence determinants of foot-and-mouth disease virus in cell culture. J. Virol. 72, 6362–6372

Barnett, P.V., Ouldridge, E. J., Rowlands, D. J., Brown, F. and Parry, N. R. (1989) Neutralising epitopes on type O foot-and-mouth disease virus. I. Identification and characterisation of three functionally independent, conformational sites. J. Gen. Virol. 70, 1483–1491

Barton, D.J., O'Donnell, B.J. and Flanegan, J.B. (2001) 5′ cloverleaf in poliovirus RNA is a *cis*-acting replication element required for negative-strand synthesis. EMBO J. 20, 1439–48

Baxt, B. (1987) Effect of lysomotropic compounds on early events in foot-and-mouth disease virus replication. Virus Res. 7, 257–271

Baxt, B. and Becker, Y. (1990) The effect of peptides containing the arginine-glycine-aspartic acid sequence on the adsorption of foot-and-mouth disease virus to tissue culture cells. Virus Genes 4, 73–83

Baxt, B., Vakharia, V., Moore, D. M., Franke, A. J. and Morgan, D. O. (1989) Analysis of neutralising antigenic sites on the surface of type A12 foot-and-mouth disease virus. J. Virol. 63, 2143–2151

Beard, C.W. and Mason, P.W. (2000) Genetic determinants of altered virulence of Taiwanese foot-and-mouth disease virus. J Virol 74, 987–91

Beckman, M.T. and Kirkegaard, K. (1998) Site size of cooperative single-stranded RNA binding by poliovirus RNA- dependent RNA polymerase. J. Biol. Chem. 273, 6724–30

Belsham, G., Abrams, C.C., King, A.M.Q., Roosien, J. and Vlak, J.M. (1991) Myristoylation of foot-and-mouth disease virus capsid precursors is independent of other viral proteins and occurs in both mammalian and insect cells. J. Gen. Virol. 72, 747–751

Belsham, G.J., McInerney, G.M. and Ross-Smith, N. (2000) Foot-and-mouth disease virus 3C protease induces cleavage of translation initiation factors eIF4A and eIF4G within infected cells. J. Virol. 74, 272–80

Bergmann, E.M., Mosimann, S.C., Chernaia, M.M., Malcolm, B.A. and James, M.N.G. (1997) The refined crystal structure of the 3C gene product from hepatitis A virus: specific proteinase activity and RNA recognition. J. Virol. 71, 2436–2448

Berinstein, A., Roivainen, M., Hovi, T., Mason, P. W. and Baxt, B. (1995) Antibodies to the vitronectin receptor (integrin $\alpha v\beta 3$) inhibit binding and infection of foot-and-mouth disease virus to cultured cells. J. Virol. 69, 2664–2666

Berman, H.M., Westbrook, J., Feng, Z., Gilliland, G., Bhat, T.N., Weissig, H., Shindyalov, I.N. and Bourne P.E. (2000) The Protein Data Bank. Nucleic Acids Res. 28,235–242

Bernfield, M., Gotte, M., Park, P. W., Reizes, O., Fitzgerald, M. L., Lincecum, J. and Zako, M. (1999) Functions of cell surface heparan sulfate proteoglycans. Annu. Rev. Biochem. 68, 729–777

Boege, U., Kobasa, D., Onodera, S., Parks, G., Palmenberg, A. and Scraba, D. (1991) Characterisation of mengo virus neutralisation epitopes. Virology 181, 1–13

Brown, F., and Cartwright, B. (1961) Dissociation of foot-and-mouth disease virus into its nucleic acid and protein components. Nature (London). 192, 1163–1164

Carrillo, E. C., Giachetti, C. and Campos, R. H. (1984) Effect of lysomotropic agents on the foot-and-mouth disease virus replication. Virology 135, 542–545

Chen, W-J., Goldstein, J. L. and Brown, M. S. (1990) NPXY, a sequence often found in cytoplasmic tails, is required for coated pit-mediated internalization of the low density lipoprotein receptor. J. Biol. Chem. 265, 3116–3123

Chow, M., Newman, J., Filman, D., Hogle, J.R., Rowlands, D.J. and Brown, F. (1987) Myristylation of picornavirus capsid protein VP4 and its structural significance. Nature (London) 327, 482–486

Cockburn, J.J.B., Abrescia N.G.A., Grimes, J.M., Sutton, G.C., Diprose, J.M., Benevides, J., Thomas Jnr, G., Bamford, J.K.H., Bamford, D.H., Stuart, D.I. (2004) A biological membrane interacting with protein and DNA visualized in bacteriophage PRD1 at high resolution. (In preparation)

Crowther, J.R., Farias, S., Carpenter, W.C. and Samuel, A.R. (1993) Identification of a fifth neutralizable site on type O foot-and-mouth disease virus following characterisation of single and quintuple monoclonal antibody escape mutants. J. Gen.Virol. 74, 1547–1553

Curry, S., Abrams, C.C., Fry, E., Crowther, J.C., Belsham, G.J., Stuart, D.I. and King, A.M.Q. (1995) Viral RNA modulates the acid sensitivity of foot-and-mouth disease virus capsid. J. Virol. 69, 430–438

Curry, S., Fry, E., Blakemore, B., Abu-Ghazaleh, R., Jackson, T., King, A., Lea, S., Newman, J., Rowlands, D. and Stuart, D. (1996) Perturbations in the surface structure of A22 Iraq foot-and-mouth disease virus accompanying coupled changes in host cell specificity and antigenicity. Structure 4, 135–145

Curry, S., Fry, E., Blakemore, W., Abu-Ghazaleh, R., Jackson, T., King, A., Lea, S., Newman, J. and Stuart, D. (1997) Dissecting the roles of VP0 cleavage and RNA packaging in picornavirus capsid stabilization: the structure of empty capsids of foot-and-mouth disease virus. J. Virol. 71, 9743–9752

Donnelly, M.L., Gani, D., Flint, M., Monaghan, S. and Ryan, M.D. (1997) The cleavage activities of aphthovirus and cardiovirus 2A proteins. J. Gen. Virol. 78, 13–21

Donnelly, M. L., Luke, G., Mehrotra, A., Li, X., Hughes, L.e.., Gani, D. and Ryan, M.D. (2001). Analysis of the aphthovirus 2A/2B polyprotein 'cleavage' mechanism indicates not a proteolytic reaction, but a novel translational effect: a putative ribosomal 'skip'. J. Gen. Virol. 82, 1013–25

Ellard, F. M., Drew, J., Blakemore, W. E., Stuart, D. I. and King, A. M. Q. (1999) Evidence for the role of His-142 of protein 1C in the acid-induced disassembly of foot-and mouth disease virus capsids. J. Gen. Virol. 80, 1911–1918

Escarmis, C., Carrillo, E. C., Ferrer, M., Arriaza, J. F. G., Lopez, N., Tami, C., Verdaguer, N. Domingo, E. and Franze-Fernandez, M. T. (1998) Rapid selection in modified BHK-21 cells of a foot-and-mouth disease virus variant showing alterations in cell tropism. J. Virol. 72, 10171–10179

Esnouf, R. M. (1997) An extensively modified version of MolScript which includes greatly enhanced colouring capabilities. J. Mol. Graphics 15, 132–134

Falk, M.M., Sobrino, F. and Beck, E. (1992) VPg gene amplification correlates with infective particle formation in foot-and-mouth disease virus. J. Virol. 66, 2251–60

Filman, D.J., Syed, R., Chow, M., Macadam, A.J., Minor, P.D. and Hogle, J.M. (1989) Structural factors that control conformational transitions and serotype specificity in type 3 poliovirus. EMBO J. 8, 1567–1579

Forss, S. and Schaller, H. (1982) A tandem repeat gene in a picornavirus. Nucleic Acids Res 10, 6441–50

Fox, G., Parry, N.R., Barnett, P.V., McGinn, B., Rowlands, D.J. and Brown, F. (1989) Cell attachment site on foot-and-mouth disease virus includes the amino acid sequence RGD (Arginine-Glycine-Aspartic Acid). J. Gen. Virol. 70, 625–637

Fox, G., Stuart, D., Acharya, K.R., Fry, E., Rowlands, D.J. and Brown, F. (1987) Crystallisation and preliminary X-ray diffraction analysis of foot-and-mouth disease virus. J. Mol. Biol. 196, 591–597

Fry, E., Lea, S.M., Jackson, T., Newman, J.W.I., Ellard, F.M., Blakemore, W.E., Abu-Ghazaleh, R., Samuel, A., King, A.M.Q. and Stuart, D.I. (1999) The structure and function of a foot-and-mouth disease virus-oligosaccharide receptor complex. EMBO J. 18, 543–554

Fry, E., Logan, D., Abu-Ghazaleh, R., Blakemore, W., Curry, S., Jackson, T., Lea, S., Lewis, R., Newman, J., Parry, N., Rowlands, D., King, A. and Stuart, D. (1991) Molecular studies on the structure of foot-and-mouth disease virus. In: P. Goodenough (Ed), Protein Engineering, pp. 71–80. CPL Press, Berkshire, UK

Fry, E., Logan, D., Acharya, R., Fox, G., Rowlands, D., Brown, F. and Stuart, D. (1990) Architecture and topography of an aphthovirus. Semin. Virol. 1, 439–451

Giancotti, F. G., and Ruoslahti, E. (1999) Integrin signalling. Science 285, 1028–1032

Grubman, M.J., Zellner, M., Bablanian, G., Mason, P.W. and Piccone, M.E. (1995) Identification of the active-site residues of the 3C proteinase of foot- and-mouth disease virus. Virology 213, 581-9

Guarné, A., Hampoelz, B., Glaser, W., Carpena, X., Tormo, J., Fita, I. and Skern, T. (2000) Structural and biochemical features distinguish the foot-and-mouth disease virus leader proteinase from other papain-like enzymes. J. Mol. Biol. 302, 1227–1240

Hansen, J.L., Long, A.M. and Schultz, S.C. (1997) Structure of the RNA-dependent RNA polymerase of poliovirus. Structure 5, 1109–22

Hendry, E., Hatanaka, H., Fry, E., Smyth, M., Tate, J., Stanway, G., Santti, J., Maaronen, M., Hyypia, T. and Stuart, D. (1999) The crystal structure of coxsackievirus A9: new insights into the uncoating mechanisms of enteroviruses. Structure 7, 1527–1538

Herold, J. and Andino, R. (2001) Poliovirus RNA replication requires genome circularization through a protein-protein bridge. Mol. Cell 7, 581–91

Hewat, E.A., Verdaguer, N., Fita, I., Blakemore, W., Brookes, S., King, A., Newman, J., Domingo, E., Mateu, M.G. and Stuart, D.I. (1997) Structure of the complex of an Fab fragment of a neutralizing antibody with foot-and-mouth disease virus: Positioning of a highly mobile antigenic loop. EMBO J. 16, 1492–1500

Hinton, T.M., Ross-Smith, N., Warner, S., Belsham, G.J. and Crabb, B.S. (2002) Conservation of L and 3C proteinase activities across distantly related aphthoviruses. J. Gen. Virol. 83, 3111–3121

Hobson, S.D., Rosenblum, E.S., Richards, O.C., Richmond, K., Kirkegaard, K. and Schultz, S.C. (2001) Oligomeric structures of poliovirus polymerase are important for function. EMBO J. 20, 1153–63

Hogle, J. M. 2002. Poliovirus cell entry: common structural themes in viral cell entry pathways. Annu. Rev. Microbiol. 56:677–702

Hogle, J.M., Chow, M. and Filman, D.J. (1985) Three-dimensional structure of poliovirus at 2.9 Å resolution. Science 229, 1358–1365

Hogle, J.M. and Filman, D.J. (1989) The antigenic structure of poliovirus. Phil. Trans. R. Soc. Lond. B 323, 467–478

Jackson, T., Blakemore, W., Newman, J., Knowles, N.J., Mould, A.P., Humphries, M.J. and King, A.M.Q. (2000a) Foot-and-mouth disease virus is a ligand for the high-affinity binding conformation of integrin $\alpha5\beta1$: influence of the leucine residue within the RGDL motif on selectivity of integrin binding. J. Gen. Virol. 81, 1383–1391

Jackson, T., Ellard, F.M., Abu Ghazaleh, R., Brookes, S.M., Blakemore, W.E., Corteyn, A.H., Stuart, D.I., Newman, J.W.I. and King, A.M.Q. (1996) Efficient infection of cells in culture by type O foot-and-mouth disease virus requires binding to cell surface heparan sulfate. J. Virol. 70(8), 5282–5287

Jackson, T., Mould, A. P., Sheppard, D., Denyer, M. and King, A. (2002) Integrin $\alpha v\beta1$ is a receptor for foot-and-mouth disease virus. J. Virol. 76, 935–941

Jackson, T., Sharma, A., Abu-Ghazaleh, R., Blakemore, W., Ellard, E., Simmons, D.L., Stuart, D.I., Newman, J.W.I. and King, A.M.Q. (1997) Arginine-glycine-aspartic acid-specific binding by foot-and-mouth disease viruses to the purified integrin $\alpha v\beta3$ in vitro. J. Virol. 71, 8357–8361

Jackson, T., Sheppard, D., Denyer, M., Blakemore, W. and King, A. (2000b) The epithelial integrin $\alpha v\beta 6$ is a receptor for foot-and-mouth disease virus. J. Virol. 74, 4949–4956

Kitson, J.D.A., McCahon, D. and Belsham, G.J. (1990) Sequence analysis of monoclonal antibody resistant mutants of type O foot-and-mouth disease virus: evidence for involvement of the three surface exposed capsid proteins in four antigenic sites. Virology 179, 26–34

Kjellen, L., and Lindahl, U. (1991) Proteoglycans— structures and interactions. Annu. Rev. Biochem. 60, 443–475

Kraft, S., Diefenbach, B., Mehta, R., Jonczyk, A., Luckenbach, A. and Goodman, S. L. (1999) Definition of an unexpected ligand recognition motif for $\alpha v\beta 6$ integrin. J. Biol. Chem. 274, 1979–1985

Lea, S., Abu-Ghazaleh, R., Blakemore, W., Curry, S., Fry, E., Jackson, T., King, A., Logan, D., Newman, J. and Stuart, D. (1995) Structural comparison of two strains of foot-and-mouth disease virus subtype O1 and a laboratory antigenic variant, G67. Structure 3, 571–580

Lea, S., Hernandez, J., Blakemore, W., Brocchi, E., Curry, S., Domingo, E., Fry, E., Abu-Ghazaleh, R., King, A., Newman, J., Stuart, D. and Mateu, M.G. (1994) The structure and antigenicity of a type C foot-and-mouth disease virus. Structure 2, 123–139

Lewis, J.K., Bothner, B., Smith, T.J. and Siuzdak, G. (1998) Antiviral agent blocks breathing of the common cold virus. Proc. Natl. Acad. Sci. USA 95, 6774–6778

Liebermann, HT., Dolling, R., Schmidt, D. and Thalmann, G. (1991) RGD-containing peptides of VP1 of foot-and-mouth disease virus (FMDV) prevent virus infection in vitro. Acta Virol. 35, 90–93

Loeffler, F. and Frosch, P. (1897) Summarischer Bericht über der Ergebnisse der Untersuchungen zur Erforschung der Maul- und Klauenseuche. Zentralblatt Bakt. Abt. Orig. 22, 257–259

Logan, D., Abu-Ghazaleh, R., Blakemore, W., Curry, S., Jackson, T., King, A., Lea, S., Lewis, R., Newman, J., Parry, N., Rowlands, D., Stuart, D. and Fry, E. (1993) Structure of a major immunogenic site on foot-and-mouth disease virus. Nature (London) 362, 566–568

Luo, M., Vriend, G., Kamer, G., Minor, I., Arnold, E., Rossmann, M.G., Boegé, U., Scraba, D.G., Duke, G.M. and Palmenberg, A.C. (1987) The atomic structure of mengo virus at 3.0 Å resolution. Science 235, 182–191

Lubroth, J. and Brown, F. (1995) Identification of native foot-and-mouth disease virus non-structural protein 2C as a serological indicator to differentiate infected from vaccinated livestock. Res. Vet. Sci. 59, 70–8

Mason. P. W., Baxt, B., Brown, F., Harber, J., Murdin, A. and Wimmer, E. (1993) Antibody-complexed foot-and-mouth disease virus, but not poliovirus, can infect cells via the Fc receptor. Virology 192, 568–577

Mason, P.W., Rieder, E. and Baxt, B. (1994) RGD sequence of foot-and-mouth disease virus is essential for infecting cells via the natural receptor but can be bypassed by an antibody-dependent enhancement pathway. Proc. Natl. Acad. Sci. USA 91, 1932–6

Martinez, M. A., Verdaguer, N., Mateu, M. and Domindo, E. (1997) Evolution subverting essentiality; dispensability of the cell attachment Arg-gly-asp motif in multiply pasaged foot-and-mouth disease virus. Proc. Natl. Acad. Sci. USA 94, 6798–6802

Mateu, M. G., Luz Valero, M., Andreu, D. and Domingo, E. (1996) Systematic replacement of amino acid residues within an Arg-Gly-Asp-containing loop of foot-and-mouth disease virus and effects on cell recognition. J. Biol. Chem. 271, 12814–12819

Matthews, D.A., Smith, W.W., Ferre, R.A., Condon, B., Budahazi, G., Sisson, W., Villafranca, J.E., Janson, C.A., McElroy, H.E., Gribskov, C.L. et al. (1994) Structure of human rhinovirus 3C protease reveals a trypsin-like polypeptide fold, RNA-binding site, and means for cleaving precursor polyprotein. Cell 77, 761–71

McCahon, D., Crowther, J.R., Belsham, G.J., Kitson, J.D.A., Duchesne, M., Have, P., Meloen, R.H., Morgan, D.O. and De Simone, F. (1989) Evidence for at least four antigenic sites on type O foot-and-mouth disease virus involved in neutralisation; identification by single and multiple monoclonal antibody-resistant mutants. J. Gen. Virol. 70, 639–645

Merritt, E.A. and Murphy, M.E.P. (1994) Raster3D version 2.0: a program for photorealistic molecular graphics. Acta Cryst. D50, 869–873

Miller, L. C., Blakemore, W. E., Sheppard, D., Atakilit, A., King, A. M. Q. and Jackson. T. (2001) Role of the cytoplasmic domain of the β-subunit of integrin $\alpha v\beta 6$ in infection by foot-and mouth disease virus. J. Virol. 75, 4158–4164

Mosimann, S.C., Chernaia, M.M., Sia, S., Plotch, S. and James, M.N.G. (1997) Refined X-ray crystallographic structure of the poliovirus 3C gene product. J. Mol. Biol. 273, 1032-

Neff, S., Mason, P. W. and Baxt. B. (2000) High-efficiency utilization of the bovine integrin $\alpha v\beta 3$ as a receptor for foot-and-mouth disease virus is dependent on the bovine $\beta 3$ subunit.. J. Virol. 74, 7298–7306

Neff, S., Sa-Carvalho, D., Rieder, E., Mason, P. W., Blystone, S. D., Brown, E. J. and Baxt. B. (1998) Foot-and-mouth disease virus virulent for cattle utilizes the integrin $\alpha v\beta 3$ as its receptor. J. Virol. 72, 3587–3594

O'Donnell, V.K., Pacheco, J.M., Henry, T.M. and Mason, P.W. (2001) Subcellular distribution of the foot-and-mouth disease virus 3A protein in cells infected with viruses encoding wild-type and bovine-attenuated forms of 3A. Virology 287, 151–62

Ochoa, W.F., Kalko, S.G., Mateu, M.G., Gomes, P., Andreu, D., Domingo, E., Fita, I. and Verdaguer, N. (2000) A multiply substituted G-H loop from foot-and-mouth disease virus in complex with a neutralizing antibody. A role for water molecules. J. Gen. Virol. 81, 1495–1505

Ohad, M., Weber, I., Frangakis, A.S., Nicastro, D., Gerisch, G. and Baumeister, W. (2002) Macromolecular architecture in eukaryotic cells visualized by cryoelectron tomography. Science 298, 1209–1213

Pacheco, J.M., Henry, T.M., O'Donnell, V.K., Gregory, J.B. and Mason, P.W. (2003) Role of nonstructural proteins 3A and 3B in host range and pathogenicity of foot-and-mouth disease virus. J. Virol. 77, 13017–13027

Palmenberg, A.C., Parks, G.D., Hall, D.J., Ingraham, R.H., Seng, T.W. and Pallai, P.V. (1992) Proteolytic processing of the cardioviral P2 region: primary 2A/2B cleavage in clone-derived precursors. Virology 190, 754–62

Parry, N., Fox, G., Rowlands, D., Brown, F., Fry, E., Acharya, R. and Stuart, D. (1990) Structural and serological evidence for a novel mechanism of immune evasion in foot-and-mouth disease virus. Nature (London) 347, 569–572

Parry, N.R., Ouldridge, E.J., Barnett, P.V. and Brown, F. (1989) Neutralising epitopes of type O foot-and-mouth disease virus. II. Mapping three conformational sites with synthetic peptide reagents. J. Gen. Virol. 70, 1493–1503

Parry, N.R., Ouldridge, E.J., Barnett, P.V., Rowlands, D.J., Brown, F., Bittle, J.L., Houghten, R.A. and R.A., L. (1985) Identification of neutralizing epitopes of foot-and-mouth disease virus. In: R.A. Lerner, R.M. Chanock and F. Brown (Eds), Vaccines '85, pp. 211–216. Cold Spring Harbor Laboratory Press, Cold Spring Harbor, New York

Paul, A.V., Peters, J., Mugavero, J., Yin, J., van Boom, J.H. and Wimmer, E. (2003) Biochemical and genetic studies of the VPg uridylylation reaction catalyzed by the RNA polymerase of poliovirus. J. Virol. 77, 891–904

Petersen, J.F.W., Cherney, M.M., Liebig, H-D., Skern, T., Kuechler, E. and James, M.N.G. (1999) The structure of the 2A proteinase from a common cold virus: a proteinase responsible for the shut-off of host-cell protein synthesis. EMBO J. 18, 5463–5475

Pfaff, E., Thiel, H-J., Beck, E., Strohmaier, K. and Schaller, H. (1988) Analysis of neutralizing epitopes on foot-and-mouth disease virus. J. Virol. 62, 2033–2040

Richards, O.C. and Ehrenfeld, E. (1998) Effects of poliovirus 3AB protein on 3D polymerase-catalyzed reaction. J. Biol. Chem. 273, 12832–40

Rieder, E., Baxt, B. and Mason, P.W. (1994) Animal-derived antigenic variants of foot-and-mouth disease virus type A12 have low affinity for cells in culture. J. Virol. 68, 5296–5299

Rossmann, M.G., Arnold, E., Erickson, J.W., Frankenburger, E.A., Griffith, J.P., Hecht, H.J., Johnson, J.E., Kamer, G., Luo, M., Mosser, A.C., Rueckert, R.R., Sherry, B. and Vriend, G. (1985) Structure of a human common cold virus and functional relationship to other picornaviruses. Nature (London) 317, 145–153

Rueckert, R.R. (1996) Picornaviridae: The viruses and their replication. In: B.N. Fields (Ed), Fields Virology, Third ed., pp. 609–654. Vol. 1. 2 vols. Lippincott-Raven Publishers, Philadelphia, PA

Sa-Carvalho, D., Rieder, E., Baxt, B., Rodarte, R., Tanuri, A. and Mason, P.W. (1997) Tissue culture adaptation of foot-and-mouth disease virus selects viruses that bind to heparin and are attenuated in cattle. J. Virol. 71, 5115–5123

Sharma, A., Rao, Z., Fry, E., Booth, T., Jones, E.Y., Rowlands, D.J., Simmons, D.L. and Stuart, D.I. (1997) Specific interactions between human integrin $\alpha v\beta 3$ and chimeric hepatitis B virus core particles bearing the receptor-binding epitope of foot-and-mouth disease virus. Virology 239, 150–157

Sherry, B., Mosser, A.G., Colonno, R.J. and Ruecket. R.R. (1986) Use of monoclonal antibodies to identify four neutralisation epitopes on a common cold picornavirus, human rhinovirus 14. J. Virol. 57, 246–257

Spanchak, K., Namy, O., Brierley, I. and Gilbert, R.C. In preparation

Stave, J.W., Card, J.L., Morgan, D.O. and Vakharia, V.N. (1988) Neutralisation sites of type O1 foot-and-mouth disease virus defined by monoclonal antibodies and neutralisation-escape mutants. Virology 162, 21–29

Strauss, D.M., Glustrom, L.W. and Wuttke, D.S. (2003) Towards an understanding of the poliovirus replication complex: the solution structure of the soluble domain of the poliovirus 3A protein. J. Mol. Biol. 330, 225–234

Surovoi, A. Y., Ivanov, V. T., Chepurkin, A. V., Ivanyuschenkov, V. N. and Dryagalin, N. N. (1988) Is the Arg-Gly-Asp sequence the site for foot-and-mouth disease virus binding with cell receptor? Sov. J. Bioorg. Chem. 14, 965–968

Thomas, A.A.M., Woortmeijer, R.J., Puijk, W. and Barteling, S.J. (1988) Antigenic sites on foot-and-mouth disease virus type A_{10}. J. Virol. 62, 2782–2789

Trowbridge, I. S., Collawn, J. F. and Hopkins, C. R. (1993) Signal-dependent membrane protein trafficking in the endocytotic pathway. Annu. Rev. Cell Biol. 9, 129–161

van Vlijmen, H.W.T., Curry, S., Schaefer, M. and Karplus, M. (1998) Titration calculations of foot-and-mouth disease virus capsids and their stabilities as a function of pH. J. Mol. Biol. 275, 295–308

Verdaguer, N., Mateu, M.G., Andreu, D., Giralt, E., Domingo, E. and Fita, I. (1995) Structure of the major antigenic loop of foot-and-mouth disease virus complexed with a neutralizing antibody: direct involvement of the arg-gly-asp motif in the interaction. EMBO J. 14, 1690–1696

Verdaguer, N., Mateu, M.G., Bravo, J., Tormo, J., Giralt, E., Andreu, D., Domingo, E. and Fita, I. (1994) Crystallisation and preliminary X-ray diffraction analysis of a monoclonal FAb fragment against foot-and-mouth disease virus and of its complex with the main antigenic site peptide. Proteins 18, 201–203

Verdaguer, N., Schoehn, G., Ochoa, W.F., Fita, I., Brookes, S., King, A., Domingo, E., Mateu, M.G., Stuart, D. and Hewat, E.A. (1999) Flexibility of the major antigenic loop of foot-and-mouth disease virus bound to a Fab fragment of a neutralizing antibody: structure and neutralization. Virology 255, 260–268

Verlinden, Y., Cuconati, A., Wimmer, E. and Rombaut, B. (2000) Cell-free synthesis of poliovirus: 14S subunits are the key intermediates in the encapsidation of poliovirus RNA. J. Gen. Virol. 81, 2751–4

Wild, T., Burroughs, N. and Brown, F. (1969) Surface structure of foot-and-mouth disease virus. J. Gen. Virol. 4, 313–320

Wistow, G., Turnell, B., Summers, L., Slingsby, C., Moss, D., Miller, L., Lindley, P. and Blundell, T. (1983) X-ray analysis of the eye lens protein g-II crystallin at 1.9 Å resolution. J. Mol. Biol. 170, 175–202

Xiang, W., Cuconati, A., Hope, D., Kirkegaard, K. and Wimmer, E. (1998) Complete protein linkage map of poliovirus P3 proteins: interaction of polymerase 3Dpol with VPg and with genetic variants of 3AB. J. Virol. 72, 6732–41

Xiang, W., Harris, K.S., Alexander, L. and Wimmer, E. (1995) Interaction between the 5′-terminal cloverleaf and 3AB/3CDpro of poliovirus is essential for RNA replication. J. Virol. 69, 3658–67

Xie, Q.-C., McCahon, D. Crowther, J.R., Belsham, G.J. and Mc Cullough, K. C. (1987) Neutralisation of foot-and-mouth disease virus can be mediated through any of at least three separate antigenic sites. J. Gen. Virol. 68, 1637–1647

Xiong, J-P., Stehle, T., Diefenbach, B., Zhang, R., Dunker, R., Scott, D.L., Joachimiak, A., Goodman, S.L., Arminaout, M.A. (2001) Crystal structure of the extracellular segment of the integrin alphavbeta3. Science 294, 339–345

CTMI (2005) 288:103–131

Natural and Vaccine Induced Immunity to FMD

T. R. Doel

Merial Animal Health Ltd., Ash Road, Pirbright, Woking, Surrey, GU24 0NQ, UK
tim.doel@merial.com

Abstract A brief overview of the foot-and-mouth disease (FMD) literature over the last 100 years will give the impression that a great deal is known about the immune response of livestock to infection and vaccination. At the practical level, this is indeed the case and our knowledge is more than adequate in relation to the production and supply of potent vaccines for the control of the disease. The deficiencies in our understanding of the immune response are at the fundamental level and, arguably, stand in the way of its rational manipulation to achieve goals such as life-long immunity conferred by vaccination. Most of the research activity to date has focused on T cell dependency of the immune response of livestock and important B (and probably T) cell epitopes and has been used by researchers to design highly sophisticated novel vaccines and delivery systems. None of these, to the author's knowledge, exceeds the potency obtained with a good commercial vaccine. Although it is not yet possible to see a clear direction for the development of improved formulations, it is important to reflect on our current knowledge of natural and vaccine-induced immunity and some of the issues surrounding modern inactivated FMD vaccines. This process will perhaps help to discriminate the fact from the fiction and serve to focus on precisely what is needed or desirable for improved products.

1 Natural Immunity

It will be apparent in the following review that much of what we know of natural *or* vaccine-induced immunity against foot-and-mouth disease (FMD) is based on studies with both infected *and* vaccinated animals. A practical consequence for the reader is that issues such as maternally derived immunity are reviewed primarily within the section on vaccine induced immunity because of its considerable relevance to field vaccination campaigns. Fortunately, it would appear that extrapolation of data from the vaccinated state to the infected state and vice versa is frequently found to be legitimate and, in fact, largely inevitable considering the difficulty of conducting significant experiments with infected animals. In addition, much of the work over the last century has been preoccupied with cattle because of their relative economic importance and only relatively recently have workers examined the immune responses of pigs and, to a lesser extent, sheep.

1.1 Cattle

Natural immunity of cattle following recovery from FMD infection is sometimes used as the benchmark by those workers who seek to improve on conventional inactivated whole virus vaccines.

According to the popular definition of the serotypes of FMD, cattle which have recovered from infection with one of the seven serotypes are not immune to the other six. However, it is not commonly known that further rounds of infection with other serotypes may result in less severe clinical responses or even protection (Cottral and Gailiunas 1972). In the work referenced, one animal was completely resistant to infection by a fourth serotype, two animals were completely resistant to a fifth serotype and one animal was completely resistant to a seventh serotype. The authors reported that the virus cross-neutralization titres of the cattle sera were consistent with the protection observed.

In the case of protection against homologous virus challenge, high-quality immunity, as conferred by a recent infection (6 months previously) appears capable of preventing development of any clinical signs of disease regardless of whether virus is administered by needle or contact challenge (T.R. Doel, unpublished observations). In earlier experiments (Cunliffe 1964), convalescent cattle were challenged approximately 1 year after exposure to virus and were found to be protected, although lesions developed at the inoculation sites. In fact, there is evidence that neutralizing antibody and protection may even persist for the effective lifetime of some cattle, as demonstrated with eight animals which were challenged 5.5 years after initial infection (Garland 1974). These findings were similar to those of another laboratory in which cattle were held for 4.5 years after initial infection (Cunliffe 1964) and, after challenge, one of three animals was protected. Nevertheless, it is necessary to qualify the experimental observations reported here on the grounds that the level of protection after a long convalescent period could be expected to depend on a number of factors including the serotype, and possibly subtype, of virus used in the experiment (i.e. relative virulence for cattle) and variations in the rate of decline of antibody for individual animals. Although most of the experiments involved challenge as well as serology, it is also worth mentioning that the well-recognised correlations which exist between FMDV specific antibody titres and protection have been made with sera taken from animals shortly after vaccination (usually 21 to 28 days) and it would be premature to assume that the antibodies in long-term convalescent animals would be qualitatively equivalent.

1.2 Sheep and Goats

Despite the recognised importance of sheep and, almost certainly, goats in the epidemiology of FMD, relatively little is known of their immune response to infection. Virus-neutralising antibodies appear 60 h after inoculation of sheep reaching a maximum titre around 10 dpi. Titres decrease slightly after day 10 but generally remain at a plateau for at least 147 days (Dellers and Hyde 1964). The carrier state may develop in sheep and goats, and virus has been recovered in oesopharyngeal specimens from 1 to 9 months and more than 1 month, respectively (reviewed by Cottral 1969).

1.3
Pigs

The immune response of pigs to infection with FMD has been studied by a number of workers. According to Cunliffe (1962), serum neutralising antibody titres initially rose to peak levels between 7 and 10 dpi and decreased thereafter 12-fold until they reached a relatively stable plateau at 28 days until the end of the experiment (128 dpi). When five convalescent pigs were challenged by contact with a homologous virus-infected pig at 128 days after the first cycle of infection, only one developed disease. These results differ slightly from the experiments of other workers (Gomes 1977; McKercher and Giordano 1967) in which approximately 50% of the animals succumbed to challenge 3 to 6 months after the first exposure to virus. It would seem, therefore, that the duration of immunity of convalescent pigs is significantly shorter than that of cattle. This may owe much to the important observation that pigs do *not* become persistently infected. The mucosal immune response of pigs also shows an interesting difference to that of cattle in that the neutralising antibody titres of the nasal fluid of pigs are very similar to those of serum throughout a period of 50 days following exposure to virus (Francis and Black 1983), suggesting a more significant role for mucosal immunity in this species.

1.4
Serum Antibody Response

The crucial role of antibody in protection against FMD is supported by both general and FMD-specific work over the last century. In a recent review on neutralizing antiviral antibody responses, Zinkernagel et al (2001) commented, 'Immune responses and protection against cytopathic virus infections are key to species survival. Without exception, protection against these agents is mediated by protective antibodies'. From the FMD perspective, the importance of virus-specific serum antibody was demonstrated by Löffler and Frosch (1897), who showed that animals could be protected by passively administered convalescent serum. In fact, immune serum was quite widely used in FMD control campaigns in the early part of the twentieth century by pioneers such as the late Charles Mérieux. This and other work led ultimately to the establishment and widespread acceptance of correlations between in vitro serum assays such as the virus-neutralising antibody test and protection of cattle, most of the data having been derived from vaccine potency trials (e.g. Ahl et al. 1990). Some of these correlations have been extended to antibodies measured by ELISA because of the significant advantages this assay offers in terms of ease of use, speed and avoidance of the need to grow cells and live FMDV. It is vital to remember, however, that the population of antibodies measured by ELISA will depend greatly on the methodology and reagents used and will not necessarily reflect the titres of important (i.e. protective) antibodies in a sample. Zinkernagel et al. (2001) considered that it was '...mandatory to assess the protective capacity of antibodies by in vivo adoptive transfer experiments or by in vitro neutralization assays, rather than measuring antigen binding in an ELISA'.

There has been some dispute within the FMD scientific community on the absolute relevance of virus-neutralising antibody in the serum of infected or vaccinated animals, based on occasional results from potency tests in which animals were or were not protected, contrary to the prediction from their neutralising antibody titres (Ahl et al 1990). However, it is vital to recognise a fundamental flaw in much of the debate over the apparent occasional failure of neutralising antibody titres to correlate with protection against live virus challenge. First, there is an understandable preference to accept the results of the challenge test, largely because of its apparent practical significance but also because of the very long history of its use. By default, therefore, the in vitro test may be perceived as deficient in some way and 'fail to correlate'. However, the Ph.Eur live virus challenge test, which uses three groups of five cattle, has a very high intrinsic statistical variability (44% to 220% of the 'real value' without reference to the statistical error due to the biology of the challenge process and virus- and species-specific considerations; Hendriksen 1988) and it is, therefore, highly improbable that any test, serological or otherwise, would correlate at all times. In fact, it may come as something of a surprise, considering the long history of FMD vaccines, that we do not know the observed between-test variation of any of the challenge tests used for testing of FMD vaccines. Furthermore, it has been suggested by one of the pioneers of the three Rs (refinement, reduction, replacement of animal tests) that tests such as the Ph.Eur cattle potency test for FMD vaccines might not have been acceptable to regulators if, historically, it had been developed after serological tests rather than preceding them (Russell 2000).

Because of the importance of serum antibody in protection against FMD, there have been a number of studies which have examined the classes and subclasses of virus-neutralising antibodies. Specific IgM may be detected 3 to 5 days after infection of cattle, reaching a peak between

5 and 10 days, whereas specific IgG_1 and IgG_2 appear from 4 days onward and reach maximal levels between 15 and 20 days (Abu Elzein and Crowther 1981; Cowan, 1973). As stated above, serum neutralising antibody titres may persist for many years, albeit at reduced levels compared to the peak responses. Although specific IgM does not generally persist, it has been detected as late as 6 months after infection in swine (Cowan 1973).

1.5 Mucosal Antibody Response and the 'Carrier State'

When reviewing immunity against FMD, it is inevitable that the subject of mucosal responses is raised, considering the importance of the oropharyngeal route in disease entry and development. FMDV-specific mucosal immune responses have been studied after infection of cattle. A peak of neutralising activity attributed to IgM and IgA was observed in the pharyngeal fluid 7 days after virus exposure and appeared to be the result of serum and tissue fluid leakage during the inflammatory phase of the infection (Francis et al. 1983). Between 20 and 60 dpi, the neutralising activity of the pharyngeal fluid was attributed exclusively to IgA produced at mucosal surfaces rather than by serum transudation (Francis et al. 1983). McVicar and Sutmöller (1974) reported that the virus-neutralising activities of both serum and oesophageal-pharyngeal fluid were higher in virus carriers[1] rather than non-carriers, suggesting the continued stimulation of both compartments of the immune response by the persistent infection.

In the establishment of the carrier state, it would appear that FMDV is able to infect, replicate and establish within privileged sites in the oropharynx before the immune system has an opportunity to mount an effective response. Thus the virus evades immune protective mechanisms for at least a period of months. In general, persistently infected cattle eventually 'cure' although this may take several years in some exceptional cases. The prevalence of persistence in cattle recovered from infection

[1] After recovery from infection, it is often possible to recover virus from oesophageal-pharyngeal fluid of a high proportion of apparently healthy cattle. This condition is sometimes referred to as the 'carrier state', which implies at the very least that a carrier animal is capable of transmitting virus to healthy livestock. In the author's opinion, the evidence for such transmission is scant whereas substantially more data would argue against an epidemiological role for carrier animals. In many respects, 'virus persistence' is a preferable description, but 'carrier' has become a familiar word in the FMD vocabulary and will be used primarily in this article.

offers an explanation for the long duration of immunity seen with some individuals—namely, the more or less continuous stimulation or perhaps frequent boosting of the bovine immune system by very low levels of carrier virus proteins. One flaw in the hypothesis of constant or pulsed antigenic stimulation is that a small but significant percentage of seronegative, persistently infected animals have been reported (Hedger 1968; Sutmöller 1971). A possible explanation for the absence of specific serum antibodies is that these animals were in a state of low dose or high dose tolerance, although there is no experimental evidence to support this. A second explanation is suggested by the findings of Sutmöller and colleagues (1970), who reported that a high percentage of cattle is infected with bovine enterovirus (BEV) and that FMDV-BEV hybrids, in which the FMDV RNA is encapsidated by the BEV structural proteins, are produced in the oropharynx after infection by FMD. Thus these FMDV-BEV hybrids would not be expected to replicate readily in the oropharynx or elsewhere because of the high titres of antibody against BEV but would allow the isolation of infectious FMDV in tissue culture of oesophageal-pharyngeal samples. Indeed, the authors suggested that BEV significantly modulated the development of clinical FMD in some animals. The contribution, if any, of such FMDV-BEV hybrids in the maintenance of the persistent state is not known.

Unfortunately, it remains difficult to analyse the contribution of mucosal immunity in the protection of cattle against FMD because mucosal neutralising antibody titres, when present, are accompanied by substantially higher titres of antibodies in the serum. The latter will not only protect against generalization but would probably reach mucosal surfaces because of tissue damage during the acute phase of the infection, unless mucosal antibody titres were sufficiently high enough to prevent early virus replication. One further consideration which certainly complicates any interpretation of the contribution that mucosal or serum antibodies play in the modulation and control of the persistent state is the possible involvement of cytokines. There has been something of a revival in this topic, and Zhang et al. (2002) have demonstrated that IFN-γ significantly restricts or even cures persistently infected cultures of bovine epithelial cells.

1.6
T Cell Responses

After infection, T cell responses have been examined both with respect to determining protective mechanisms other than those mediated by neutralising antibody and to defining the possible role(s) of T cells in support of FMD specific antibody production by the B cell population.

FMDV immediate and delayed-type hypersensitivity reactions have been observed with guinea-pigs (Knudsen et al. 1979) and cattle (Sharma et al. 1985) but no correlation was apparent between the responses in guinea-pigs and protection against challenge. More recent studies (Childerstone et al. 1999) attempted to characterize classic cytotoxic T cells, but these were thwarted by the problem of establishing a reproducible assay and the authors resorted to an in vitro proliferation assay. Although they were unable to conclude a significant role for cell-mediated immune responses, they raised the possibility that T cells could be important in the elimination of virus within carrier animals.

Although a few early experiments led to the suggestion that FMDV might be a T-independent antigen (Borca et al. 1986), almost all of the published work (e.g. Collen et al 1989; Collen and Doel 1990) points to T cell dependency and includes studies with cells isolated from animals after infection or vaccination with whole virus or synthetic peptides. The availability of synthetic peptides, in particular, has facilitated the identification of short sequences recognised by FMDV-specific T cell populations, often with the general aim of constructing novel vaccine antigens carrying both B and T cell epitopes.

In two separate studies with peripheral blood T cells taken from infected cattle, only low to moderate proliferative responses were observed when stimulated in vitro with virus (Collen 1991; Garcia-Valcarcel 1993), similar, in fact, to the T cell responses of a vaccinated group. This was despite the development of high levels of neutralising antibody in all animals. Furthermore, three of the cattle were shown to be completely protected, including the inoculated tongues, against homologous challenge some 6 months after recovery from the first infection (T.R. Doel, unpublished observations; Garcia-Valcarcel 1993). Some of this work was done with MHC class II typed animals and there was no evidence of MHC restricted T cell responses to whole virus (Collen 1991). Another interesting feature of these experiments was the biphasic nature of both the neutralising antibody and T cell proliferative responses, in which an initial coincident peak could be observed at 14 dpi followed by a trough at approximately 28 dpi and a subsequent gradual increase in both responses up to the end of the experiment at 56 dpi. A possible explanation for the recovery of the antibody and T cell responses after 28 dpi could be the establishment of the carrier state and the continued presence of viral antigens. Although the in vitro proliferative responses of peripheral blood T cells of once-infected cattle were generally weak, a second in vivo challenge resulted in significantly stronger in vitro T cell responses (Garcia-Valarcel 1993). Of course, the in vivo significance of 'weak' and 'strong' in vitro T cell proliferative responses is not known and must be interpreted with caution. In addition, it may be that the circulation of FMDV-specific, active T cells is limited because of their tissue tropism and because the majority are sequestered in the lymph nodes or other tissues of the lymphoreticular system. No information is available on the duration of T cell responses in FMDV-infected cattle.

1.7 Viral Antigens and Epitopes

Our knowledge of viral antigens and antigenic structures which are recognised by the immune system owes much to in vitro studies with sera from infected, but particularly, vaccinated animals and will be reviewed also within the vaccination section of this paper. However, there is no evidence to suggest that the immune response to infection is fundamentally different to that to vaccination in terms of the antigens which are recognised. It is clear that the host immune system recognises a number of large, multiprotein antigens as well as defined epitopes within those antigens and is, almost inevitably, likely to be more complex than the in vitro studies would suggest. The whole virus, often referred to as 146S on the basis of its sedimentation coefficient, is undoubtedly the most important viral antigen as shown by many experiments. Mild acid or heat treatment largely destroys the protective capacity of experimental vaccines (Doel and Chong 1982) and produces the so-called 12S particles, which contain five copies of each of three of the four structural proteins of the virus (VP1, VP2 and VP3). The remaining protein, VP4, does not appear to play any key role in the immune recognition of the whole virus. Some serotypes of FMDV produce so-called natural empty particles or 75S particles in tissue culture (Doel and Baccarini 1981) which appear to be antigenically very similar to 146S particles but lack the RNA of the virus. From in vitro studies, it is assumed that natural infection of cattle would also generate natural empty particles.

After infection of a host cell, a number of other proteins are also translated from the viral genome, and these are necessary in the replication and morphogenesis of new virus particles. These so-called non-structural proteins (NSPs) are capable of stimulating an immune response, and, for many years, antibody titres against one of them, VIAA (virus infection-associated antigen; Newman et al. 1979), were widely regarded as a reliable serological indicator of infection. Although this has been refuted by the observation of antibodies against VIAA after repeated vaccination with crude but completely inactivated FMD vaccines (Pinto and Garland 1979), antibodies against some or all of the other NSPs (3AB, 3ABC, 2C, 3A, 3B, 3D) are now considered to be a valuable tool in discriminating infected from vaccinated animals (Manual of Standards 2000) provided the vaccines are formulated with inactivated, purified FMDV. Unlike the structural proteins of the virus, the sequences of the various NSPs are highly conserved among the serotypes, offering the possibility of the immune response against them being boosted by a serotype unrelated to the previous round of infection. It is tempting to suggest that this might be the mechanism of cross-protection following rounds of infection with different serotypes observed many years ago (Cottral and Gailiunas 1972).

NSP antibody assays are also valuable in examining the carrier state in cattle. Whereas the classic approach of virus isolation from oesopharyngeal specimens may be successful for up to 2.5 years (Hedger 1968; Salt 1993), the frequency of virus recovery becomes quite sporadic and intervals as long as 100 days between successful isolations have been reported (Bergmann et al, 1993; Salt 1993). Sensitive NSP antibody assays such as the EITB test, in which serum antibodies against nitrocellulose-blotted NSPs of the virus are detected immunoenzymatically, indicate that FMDV may persist beyond the last time point at which infectivity in oesopharyngeal specimens can be detected (Bergmann et al. 1993). The NSPs used in the EITB test are 2C, 3A, 3B, 3ABC and 3D (viral RNA polymerase) and are prepared from recombinant *E. coli*.

2
Vaccine-Induced Immunity

Although it is not the subject of this chapter to consider in detail FMD vaccines and their application, it is necessary to briefly review a number of characteristics and properties in as much as they relate to induction of immunity.

2.1 FMD Vaccines and Vaccination

With modern FMD vaccines made under European standards of Good Manufacturing Practice (GMP), the virus is produced in large-scale suspension cultures of baby hamster kidney cells (BHK-21) and the clarified product is inactivated with an aziridine, binary ethyleneimine, with a double-dose, two-tank process (Bahnemann 1990). In the author's company, the inactivated antigen is concentrated by ultrafiltration and then purified[2] by industrial scale chromatography to remove extraneous proteins including the NSPs of the virus. This level of purification facilitates discrimination between vaccinated animals and those previously exposed to virus with a test such as the EITB (Bergmann et al. 1993). In the final part of the production process, the antigen(s) are mixed with appropriate buffers and adjuvant(s) to potentiate the immune response. Usually this is $Al(OH)_3$ gel and saponin or a mineral oil formulation which produces a single or complex (commonly known as double oil emulsion or DOE) emulsion. The former is used for subcutaneous administration in cattle, sheep and goats, and the latter is used for intramuscular vaccination of pigs because of their poor response to $Al(OH)_3$/saponin vaccines. Oil-based vaccines are also used with cattle and, occasionally, sheep and goats. In relation to age, the dosage for a young animal is the same as that for an adult animal and there is little evidence to suggest differences between the responses of the two age groups (Nicholls et al. 1985).

FMD vaccines frequently contain more than one serotype of virus and occasionally different strains of the same serotype. Thus the serotype barrier is considered to be inflexible as far as vaccines are concerned, although the author is not aware of any vaccine studies equivalent to those reported with consecutive rounds of infection using each of the seven serotypes (see this review; Cottral and Gailiunas 1972). For practical purposes, it is accepted that animals vaccinated with one serotype of FMD are not protected against the remaining six serotypes, and there are experimental as well as field data to support this conclusion. In one large study with cattle (Black et al. 1986), it was shown that each virus serotype within a South American trivalent vaccine (A,O and C serotypes) acted independently and neither enhanced nor depressed the serological (virus neutralisation test) responses to the other serotypes. It is

[2] Concentrated, purified antigens are particularly stable when stored at liquid nitrogen temperatures and are widely used for national and international emergency reserves (antigen banks).

equally clear that the widespread practice of multiple rounds of vaccination with bi-, tri- and sometimes quadrivalent vaccines does not prevent outbreaks due to significantly different antigenic variants of FMD. The serotype barrier for vaccine may extend partially to strains within some serotypes. For example, a number of A serotype strains such as A_{24} Cruzeiro, A_{22} and A Iran 96 are sufficiently distinct antigenically as to require vaccines against each strain. In fact, it is quite common practice in situations where vaccine is used extensively to adapt a local field strain for vaccine production so as to optimise the match between the vaccine and the local field isolates. For some strains within a serotype, where antigenic differences are less significant, 'complementation' of antibody responses may occur. Thus multiple rounds of vaccination with one strain of FMD will result in an increased likelihood of protecting against related strains (Pay 1984). This does not appear to be due to a 'broadening' of the specificity of the immune response to a particular strain but rather a simple quantitative increase in cross-reactive antibody titres. Thus the basic relationship among strains of the same serotype does not change after multiple vaccinations (Dubourget et al. 1987). The implications of this are that, as antibody titres rise and fall from vaccination to vaccination, protection against a field strain closely matched to the vaccine strain will last longer than that against a less closely related isolate.

Two of the most important factors affecting the timing and frequency of vaccinations are the presence or otherwise of maternally-derived antibodies in the young animals and the levels of disease in the region.

2.2 Serum Antibody Response

As indicated above, virus-neutralising antibody is crucial in the development of protection against FMD and many studies have been conducted to establish a correlation between the titre of antibody induced by vaccination and the level of protection against live virus challenge. Perhaps the best-known work was published by Ahl and his colleagues in 1990, in which regression equations ($\log_{10}$ antibody titre vs. $\log_{10}$ PB_{50}) were developed for each of three serotypes of FMD based on a large database of vaccine trials assembled from various FMD research and vaccine manufacturing laboratories. The equations are:

$$Y = 0.923x + 0.54 \quad \text{(A serotype)}$$

$$Y = 0.923x + 0.70 \quad \text{(O serotype)}$$

$$Y = 0.923x + 0.45 \quad \text{(C serotype)}$$

and they point to slight differences between these three serotypes in terms of the titre of antibody required to achieve protection. For example, taking an antibody titre of 1.4 $\log_{10}$ and applying these equations, the corresponding PB_{50} values calculate as 8.5, 5.7 and 10.7, respectively. It is probable that similar differences exist with the other four serotypes, but there is no evidence, circumstantial or otherwise, to suggest that this extends to within serotypes. In any case, any attempt to demonstrate such minor differences between strains and even serotypes would represent a massive logistical problem remembering that the 90% confidence limits of the challenge test due to intrinsic variation alone is of the order of 5 to 22 for a 10 PD_{50} vaccine (Hendriksen 1988). Interestingly, it is characteristic of the O serotype in general (i.e. South American as well as Asian and Middle Eastern isolates) that a higher payload of 146S antigen of the O serotype is required to ensure a level of antibody equivalent to the A and C serotypes. That is, by comparison with the A and C serotypes, the O serotype 146S particles are less effective in the induction of neutralising antibodies.

FMD-specific IgM is reported to develop between 2 and 4 days after vaccination of cattle (Abu Elzein and Crowther 1981) and may persist for a considerable time (possibly more than 80 days; cited by Collen 1991). The titres also appeared to be higher in vaccinated than infected animals without a distinct peak (Abu Elzein and Crowther 1981). IgM has been shown to be more cross-reactive than the IgG isotypes in neutralization assays (Garland 1974). IgG_1 was recorded after 4 days and increased throughout the sampling period of 40 days whereas IgG_2 developed after 9 days, with a similar profile, and tended to peak at 35 days. However, it is important to qualify these observations. First, it is to be expected that the rate of development and absolute titres of the different isotypes of antibody are almost certainly dependent on the adjuvant used [Abu Elzein and Crowther (1981) used $Al(OH)_3$/saponin]. Second, earlier studies based on discrimination between isotypes, particularly IgG_1 and IgG_2, must be treated with some caution unless highly specific anti-isotypic Mabs were used (Mulcahy et al. 1990). Whether or not the IgG_1 and IgG_2 isotypes against FMDV have different protective capacities is not clear but it is difficult to believe that there will not be slight differences at least, other characteristics such as antigenic specificity, titre and avidity being equal.

In this regard, it has been reported that FMD virus vaccines induced more IgG_1 than IgG_2 whereas several peptide vaccines generally induced more IgG_2 than IgG_1, suggesting that IgG_1 is a more relevant antibody to protection of cattle given the superiority of virus vaccines over peptide vaccines (Mulcahy et al. 1990). It should be noted that the peptide vaccines used in this study induced high levels of neutralising antibody but only low levels of protection. If bovine IgG_1 is indeed superior to IgG_2 in protection against FMD then the possibility exists that the presence of significant titres of high-affinity IgG_2 may be counterproductive by binding to virus and preventing attachment of IgG_1. The complicating factor in attempting to identify the different characteristics of antibodies which contribute to protection is that the researcher is obliged to work with a polyclonal antibody population composed of a multiplicity of antibodies varying in titre, isotype, affinity and specificity. The dissection of this population represents a formidable task. On the question of affinity, the most useful data relate to peptide vaccines, where it has been shown that high-affinity anti-peptide antibodies were more likely to be protective than low-affinity antibodies (Steward et al. 1991).

2.3 Maternally Derived Immunity

Maternally derived antibodies (MDA), as a result of either previous infection or vaccination of the dam, represent an extremely important modulator of the immune response of the young animal to vaccination. MDA specific for FMDV has been reported to persist in calves and pigs for up to 5 months and 2 months, respectively (cited by Kitching and Salt 1995). It is widely accepted on the basis of evidence from many workers (reviewed by Kitching and Salt 1995) that MDA inhibits the ability of calves to respond to FMD vaccines as measured by virus-neutralising antibody titres in serum. The mechanism for this was originally thought to be a form of antigen sequestration, but more recent evidence suggests a more direct effect of MDA on the immune system (Kitching and Salt 1995). Whether or not MDA influences negatively or otherwise the development of immunological memory after vaccination is not known but deserves attention.

Unfortunately, it seems that MDA offers a 'window of opportunity' for FMD to infect a young animal whereby waning MDA levels may be insufficient to prevent virus from establishing an infection but sufficient to inhibit the development of an adequate immune response to vaccination (Kitching and Salt 1995). During this critical period, the only measure that can be usefully employed is to ensure that young livestock are kept in isolation from other animals.

In areas of low risk of contact with FMD and the absence of MDA, vaccines prepared from purified components may be used to vaccinate cattle, pigs, sheep and goats from about 14 days of age, and we have some work to show that earlier vaccination is possible without negatively impacting the immune response or causing unacceptable reactions to the vaccine. Under the same conditions but with young animals from vaccinated dams, it is necessary to delay the initial vaccination point to about 2.5 months, which allows MDA to fall to levels which do not interfere significantly with the response to the vaccine. A second dose of vaccine is usually applied approximately 4 weeks after the first dose to cattle and those pigs and small ruminants likely to be kept beyond 6 months of age. Cattle and pigs are usually revaccinated every 6 months, and annual revaccination may be applied to sheep and goats. The presence of circulating FMD in the region requires a modification to this strategy whereby vaccination is made at 2 months rather than 2.5 months for young animals with MDA. A second dose of vaccine 4 weeks after the first dose is appropriate for all livestock, and the revaccination interval is adjusted to every 4 months for cattle and pigs and every 6 months for sheep and goats.

The timing of vaccinations as detailed above, and with particular respect to the primary course, is intended to provide a level of protection consistent with most epidemiological situations and must, therefore, be considered a compromise. Indeed, vaccine users frequently ask whether it is possible to give the second dose sooner than 3 or 4 weeks. The experiment shown in Fig. 1 demonstrates some of the variables in the first year of vaccination. It can be seen that a primary course of two vaccinations separated by 2 weeks gives virtually no 'boosting' of the serological response whereas increasing the interval to 4 weeks or, particularly, 2 months gives stronger secondary responses. Equally, the 2-month interval gives excellent titres of antibody for a sustained period in contrast to the primary courses of vaccination where the intervals were shorter. Unfortunately, the delay in giving the second dose 2 months after the first provides a period of vulnerability when the antibody titres from the single vaccination are relatively low and livestock would be at risk to high levels of circulating field viruses. Thus the second dose is invariably administered at 3 to 4 weeks so that the susceptibility 'window' of the livestock is minimised.

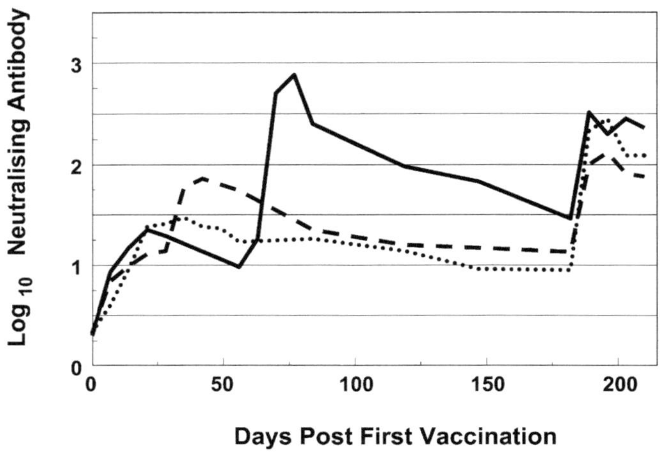

Fig. 1 Groups of five cattle were vaccinated with three doses of aluminium hydroxide/saponin vaccine over a period of approximately 6 months. Group 1 (*dotted line*) was vaccinated at day 0, day 14 and 6 months. Group 2 (*dashed line*) was vaccinated at day 0, day 28 and 6 months. Group 3 (*solid line*) was vaccinated at day 0, 2 months and 6 months. Virus-neutralising antibody titres were measured with the individual sera and the results expressed as mean titres. Vaccine strain was A22 Iraq 24/64

2.4 Early Immunity

A number of studies have demonstrated the protection of cattle, sheep and pigs within a few days of vaccination (Graves et al. 1968; Sellers and Herniman 1974; Doel et al. 1994; Barnett et al. 2002). In the author's experiments, cattle vaccinated 4 days previously with high-potency vaccines were protected against a severe challenge by contact with viraemic pigs. When the same approach was adopted with vaccinated pigs, early immunity was not demonstrated, due probably to the fact that the control pigs were housed with vaccinated animals and created conditions of overwhelming challenge. This problem has been described previously (De Leeuw et al. 1979), and the massive titres of virus shed by infected pigs appear capable of overwhelming even well-vaccinated pigs, greatly complicating the design and interpretation of any challenge work with this species. Other authors have demonstrated early protection of vaccinated pigs by isolating challenge controls and removing infected pigs as clinical disease became apparent (Salt et al. 1998). All of the recent work has been done with so-called high-potency emergency vaccines, but it should be emphasised that most of the earlier studies with cattle and pigs were certainly conducted with 'routine' vaccine preparations, suggesting that the phenomenon would occur with all FMD vaccines having an adequate potency (3 PD_{50} or greater).[3] The mechanism(s) involved in early resistance to virulent challenge is not known, but the general absence of significant titres of virus-neutralising antibodies soon after vaccination points to an alternative process. Barnett et al. (2002) have studied early protection in pigs and found elevated levels of the cytokines IL-6, IL-8 and IL-12. They attributed the cytokine profiles of the challenged pigs to the induction of monocytic cell activity and a corresponding mobilisation of the innate immune system. This prompts the question of whether or not very early immunity is serotype specific or even dependent on the presence of FMD viral antigens.

2.5 Mucosal Antibody Response and the 'Carrier State'

Mucosal immune responses of cattle due to parenteral vaccination are very different to those seen with infection. After oil emulsion vaccine, ELISA of pharyngeal fluid showed a rapid development of IgG but more delayed IgM response (Francis et al. 1983). Vaccinations at approximately 60-day intervals produced a series of IgM and IgG peaks, but at no time was IgA apparent, in contrast to parallel experiments with infected animals. Neutralising activity in the pharyngeal fluid did not appear to coincide with the first appearance of IgG but rather peaked at the second vaccination timepoint (63 days). In earlier studies using once- or twice-vaccinated animals, Garland (1974) reported that neutralisation in the pharyngeal fluid was principally associated with the IgG_1 class, and it is interesting to note that this was the class of antibody which transferred most effectively from the blood of passively immunised cattle to the mucosae of the oropharynx. There was also evidence that this selective transfer may have been assisted by increased vascular permeability due

[3] There is often confusion over the potency values given in the literature. In an older version of the cattle challenge test in which different doses of vaccine were prepared by dilution with adjuvant, the resultant PD_{50} value is approximately the square of the value obtained with another version in which both adjuvant and antigen were diluted by an inert diluent (PB_{50}). The older test was favoured largely by British workers and, therefore, publications by them must be interpreted in this light. Thus Sellers and Herniman (1974) refer to a cattle vaccine of >93 PD_{50} being used in pigs but this would only represent about 9 PB_{50} if tested by the current Ph.Eur. procedure which is now based on fractional doses rather than dilution with an inert buffer.

to hypersensitivity reactions. Clearly, the possible presence of serum-derived antibodies in the pharyngeal fluid due to transudation complicates the interpretation of such studies (Francis et al. 1983).

Neutralization due to IgA as well as IgG was detected in the secretions of a steer after a third dose of $Al(OH)_3$/saponin vaccine approximately 1 year after first vaccination (Garland 1974). This work perhaps indicates that the adjuvant used or the frequency and timing of vaccine application could be an important factor in the induction of mucosal IgA responses.

From these studies, it would be concluded that one or two rounds of conventional FMD vaccine do not normally induce significant levels of IgA in the oropharynx, whereas serum-derived neutralising IgG_1 may reach the mucosae. Multiple rounds of vaccination with a conventional product or, perhaps, only several applications of a highly potent vaccine could result in the appearance of secretory IgA. In this regard, IgA was demonstrated in the nasal fluid of four sheep 9 days after vaccination with six times the normal dose of aqueous vaccine (Gibson et al. 1984).

Primovaccination of cattle does not prevent the establishment of the 'carrier' state. Despite this observation, there is both direct and circumstantial evidence to suggest that high-quality immunity is able to significantly limit and possibly 'cure' persistently infected cattle. At the field level, Anderson and colleagues (Anderson et al. 1976) showed a much lower prevalence of the carrier state in routinely vaccinated herds of cattle although their preferred explanation was that vaccination simply reduced the levels (and hence threat) of circulating virus within the herd. More direct evidence is given by some experiments of the author (Doel et al. 1994) in which groups of cattle were vaccinated with a high-potency oil or aqueous vaccine and challenged at different time intervals after the vaccination. With the exception of the unvaccinated controls, all of the cattle were protected. In general, the frequency of persistently infected animals was high and did not appear to be significantly different from previous reports. However, it was clear that there was a decreased probability of isolating virus from the oropharyngeal fluid with increasing interval between vaccination and challenge. This observation was independently confirmed by Sorensen et al. (1998), using sera from the author's experiment in a 3D, 3AB, 3ABC blocking ELISA. In this work, it was demonstrated that seroconversion against 3A and 3ABC only occurred in one out of five cattle challenged with O Lausanne virus at day 16 dpv whereas the frequency of NSP antibody seroconversion was greater when the interval between vaccination and challenge was less. The conclusions from the O Lausanne challenge study were supported by an extension of the experiment in which the cattle were randomised and revaccinated with a different serotype vaccine (C Oberbayern) approximately 4 months after the start of the first part of the experiment. Once again, with the exception of the unvaccinated controls, all of the cattle were protected when challenged with C Oberbayern virus. However, the percentage of carrier animals was very much lower than with the O Lausanne phase of the experiment and recovery of virus from oropharyngeal fluid considerably less frequent. The results with the C Oberbayern phase of the experiment could be considered consistent with the boosting of a cross-reactive immunity against the highly conserved NSPs of the virus. Clearly, it would be valuable to carry out a substantial experiment using repeated administrations of a modern potent vaccine to determine whether high-quality immunity could prevent or 'cure' the carrier state.

Neutralising antibody responses were observed in pig nasal fluid within 3 to 7 days of injection of an oil vaccine (water-in-oil or DOE), but the titres thereafter were relatively unaffected by subsequent vaccinations at 56 and 117 dpv (Francis and Black 1983). In contrast, the serum neutralising antibody titres increased after each vaccination, the water-in-oil preparation being the most effective. Unfortunately, the authors were unable to discriminate the antibody classes responsible for the neutralising activity in the pharyngeal fluid. An important conclusion from the paper by Francis and Black is that the magnitude of the mucosal immune response in the pig following vaccination may partly explain why attempts to establish a correlation between protection and serum antibody levels in this species have not so far proven successful, in contrast to cattle. Equally, the possibility that the pig mucosal immune system is more 'responsive' to FMD infection or parenteral vaccination than that of cattle could provide a partial explanation for the fact that pigs do not become persistently infected.

2.6 The Secondary Immune Response and Duration of Immunity

One of the more striking variations in the immune response to FMD vaccines is the extent to which it is boosted by subsequent vaccinations. The increase in neutralising antibody titre may often exceed 1.5 $\log_{10}$ (Pay 1984) whereas little or no boosting of the immune response is apparent in the data from some other reports (Francis et al. 1983; Rocha et al. 1983), suggesting that the initial and subsequent vaccinations did not stimulate the development of immunological memory. This prompts the question of what variables impact on the secondary response and, by extrapolation, on the duration of protective antibody titres. Antigen payload has been studied by Rweyemamu and colleagues (Rweyemamu et al. 1984) using $Al(OH)_3$/saponin FMD vaccines in cattle containing between 7 and 329,760 ng of 146S per dose. These authors concluded that the primary response was dose dependent but that there was no evidence of low dose tolerance or high dose immunological paralysis. Whereas the highest primary antigen doses produced the highest antibody titres, the secondary responses were generally superior following primary doses of 42 or 254 ng virus. Thus the difference between the primary and secondary antibody responses was most marked when low antigen doses were used initially (Black et al. 1984). The suggested explanation for this was that high initial antibody titres limited anamnestic responses either by sequestering antigen or by regulating the immune response more directly.

Another variable which impacts on the magnitude of the secondary antibody response is the interval between the first and second vaccinations. In Fig. 1, it can be seen that the magnitude of the response to a second vaccination at 14 days was almost non-existent. Whereas a 28-day interval was more satisfactory, extending the interval to 2 months or longer resulted in impressive secondary antibody responses. This work and the 146S dose studies (Black et al. 1984; Rweyemamu et al. 1984) are very much consistent with classic studies with model antigens (Siskind and Benacerraf 1969) in which an optimum dose of an antigen exists with respect to rate of development, peak titres and persistence of high-affinity antibody, and boosting of the immune response is favoured by an increased time interval between the two immunisations.

It is probably a general perception that the duration of protective immunity after FMD vaccination is short-lived. Certainly, this is the case after a single immunisation and significant antibody titres usually wane after several months, although this will depend on the type and quality of vaccine used and the possible interference from maternally acquired antibodies. It is also frequently assumed that oil-adjuvanted FMD vaccines give a more gradual development of the antibody response compared to aluminium hydroxide and saponin formulations, and although this may be seen with some oil formulations (McKercher et al. 1975a, b), it is not always the case and primary antibody responses often peak around 21 to 28 days after vaccination.

The duration of protective antibody titres beyond the first vaccination is, equally, a controversial area. Types of vaccination programme used are heavily influenced by epidemiological circumstances and invariably are based on revaccination every 4, 6 or 12 months. It has been claimed that the frequency of immunisation with $Al(OH)_3$/saponin vaccines can be reduced with oil-adjuvanted vaccines (Barteling and Vreeswijk, 1991) because of the more prolonged antibody response observed with some oil-based formulations. This was suggested by the work of Gomes et al. (1980). In their experiment, 30 cattle which had been vaccinated three times at 6-month intervals with oil-adjuvanted vaccines (water-in-oil single oil formulations) were challenged 13 months after the last vaccination. A high level of protection was predicted according to the pre-challenge antibody levels and, indeed, 29 animals were fully protected. In contrast, there is published data which indicate that the duration of immunity after repeated application of $Al(OH)_3$/saponin vaccines can be very long (Fish et al. 1969) and not differ significantly from vaccination with oil vaccines (Kitching 1997). The work of Fish and colleagues with Frenkel-type $Al(OH)_3$/saponin vaccines is particularly noteworthy. They showed that cattle selected from the annual vaccination programme of the Netherlands during the 1960s initially produced a peak of neutralising antibody after revaccination which lasted about 12 weeks, after which time antibody titres levelled off and remained at this plateau level for up to 44 months after primary vaccination. Interestingly, the duration of antibody response was not significantly affected by the number of re-vaccinations (2, 3, 4, 5, 6, 7 or greater) whereas the plateau heights were increased slightly each time (approximately 0.1 $\log_{10}$ per re-vaccination cycle up to 4/5 revaccinations). This impressive duration of antibody responses is well supported by a number of observations by European workers after the cessation of vaccination in 1990/1991. In particular, Remond et al. (1998) carried out a large study of French cattle and reported significant titres of FMD virus-neutralising antibody titres some 6 years after the last vaccination ($Al(OH)_3$ /saponin vaccines). This was the case even with cattle which would have only received several vaccinations at the point that vaccination was prohibited. As with the kinetics of the antibody response, it is quite clear that claims on the superiority of oil-adjuvanted formulations in this respect need to be interpreted cautiously given the very different oil formulations available internationally and the proven high quality of some $Al(OH)_3$/saponin products. Equally, it is clear that long duration of antibody responses can be achieved with relatively infrequent vaccinations.

Almost all of the information on the duration of immunity has been obtained in cattle. However, the more limited data from pigs and sheep do not appear to differ greatly with respect to the rate of antibody development and decay. In the case of sheep, it has been shown that one type of oil-adjuvanted vaccine gave improved duration of antibody titre compared with $Al(OH)_3$/saponin vaccines but this was not reflected in the protection against challenge at 9 months after vaccination (13 sheep protected/14 sheep challenged and 13 sheep protected/13 sheep challenged, respectively, McKercher et al. 1975).

2.7 T Cell Responses to Vaccination

Vaccination with $Al(OH)_3$/saponin vaccines induces a population of T-lymphocytes in bovine peripheral blood which proliferate in vitro in the presence of optimum concentrations of FMD virus (Collen and Doel 1990; Collen 1991). Whereas these proliferative responses tend to be relatively weak after only one vaccination (Garcia-Valcarcel 1993; Van Lierop et al. 1992) higher stimulation indices are obtained when cattle have been vaccinated several times. Furthermore, peripheral blood T cell responses after infection do not appear to be superior to those achieved with vaccination (Garcia-Valcarcel 1993). Although the evidence is quite limited, the duration of FMDV-specific T cell immunity has been reported as 20 months for one twice-vaccinated animal (Collen and Doel 1990). The possibility also exists that non-circulating FMDV-specific T cells are resident for even longer periods than peripheral blood T cells.

A significant feature of the bovine T cell response to FMDV vaccination or infection is the cross-reactivity among the different serotypes (Collen and Doel 1990; Collen 1991; Garcia-Valcarcel 1993; Van Lierop et al. 1992) in clear contrast to the serum antibody responses. For example, peripheral blood lymphocytes from an animal previously vaccinated with O serotype virus proliferate strongly and often equivalently when stimulated in vitro with A serotype virus (Collen and Doel 1990). This cross-reactivity was attributed to conserved epitopes within VP1, VP2 or VP3 of the different serotypes of the virus. Importantly, there does not appear to be any correlation between neutralising antibody titres and T cell proliferative responses (Van Lierop et al. 1992).

The development of in vitro T cell proliferation assays and the availability of synthetic peptides prompted investigations to locate and define T cell epitopes within the structural proteins of FMDV. Collen et al. (1991) screened a number of VP1 synthetic peptides and identified an immunodominant T cell epitope between residues 21 and 40 which was recognised by 7 of 19 virus-vaccinated cattle. Sequences containing the 141–160 region of VP1 were less effective in T cell proliferation assays.

In fact, studies from this laboratory (Collen 1991; Collen et al. 1991; Garcia-Valcarcel 1993; T.R. Doel, unpublished observations) were generally unsuccessful in generating strong T cell responses to peptides containing the VP1 141–158 sequence or showing their recognition by T cells from virus-vaccinated cattle (Collen 1991). For reasons which were not apparent, Van Lierop et al. (1992) had more success with peptide-immunised cattle although one repeatedly virus-vaccinated animal gave very little T cell proliferation in response to peptides. In subsequent studies with in vitro priming of T cells from naïve animals, Van Lierop et al. (1995) concluded that although 140 to 160 sequences of VP1 were able to elicit significant peptide-specific T cell responses, they consistently failed to prime for virus-specific T cells and to a large extent vice versa. These authors also demonstrated VP4 (20–34) as a dominant T cell epitope, which satisfied the criteria of high sequence conservation among the serotypes, induction of virus-specific T cell responses and high MHC promiscuity. The latter property is widely regarded as essential in any use of peptides as synthetic vaccines given the general and specific observation of MHC restriction of T cell responses to synthetic peptides (Glass et al. 1991 for FMDV).

T cell responses against whole virus and isolated proteins have also been demonstrated in C serotype-vaccinated inbred miniature pigs (Saiz et al. 1992), including the observation that the response was heterotypic. After two vaccinations, the T cell responses persisted for at least 1 year although the quantities of antigen required for in vitro proliferation were particularly high (40–200 μg/ml) in comparison with the cattle experiments referred to above.

2.8 Vaccine Antigens

FMD antigen harvests contain a high proportion of irrelevant cellular and growth media-derived proteins and small quantities of structural proteins and NSPs of FMDV in a very approximate ratio of 1,000 to 1 by weight of protein. The structural proteins include 146S particles and various subunits including natural empty particles (75S) and pentameric clusters of VP1, VP2 and VP3 (12S). The immunogenicity of 146S particles exceeds that of the 75S particles (found predominantly in the A serotype viruses) and the 12S particles by approximate factors of 10 and 100, respectively (Doel and Chong 1982). The integrity of the 146S particle is thus crucial to the efficacy of a vaccine and, in the author's laboratory, is the key parameter for determining antigen payload within a vaccine formulation. Within the intact 146S particle, the VP1 protein may be partially degraded by the action of some proteolytic enzymes and, depending on the virus strain, this may reduce the ability of 146S particles to induce neutralising antibodies. This appears to be a problem primarily with relatively crude virus concentrates stored at 4°C (Doel and Collen 1982), and purification procedures substantially reduce proteolytic activity.

Rweyemamu et al. (1984) examined the dose response of cattle given vaccines formulated with different 146S payloads. From their data, it was clear that protective antibody responses were induced above payloads of approximately 1 μg after a single dose of O serotype vaccine. It appeared that there was little benefit to antibody titre when payloads above 10 μg were used, and examination of the data suggests perhaps a sigmoid relationship between antigen concentration and antibody response where the straight line region is between 1 and 10 μg. As mentioned above, higher antigen payloads are required with the O serotype compared to most other FMD serotypes.

Although the immune response to FMD vaccine is serotype specific, it has been suggested that the 12S antigen of the virus is able to stimulate inter-serotypic antibody responses (Cartwright et al. 1982). However, this is difficult to reconcile with the fact that all but the purest conventional FMD vaccines probably contain a significant quantity of 12S (approximately 50% of the CF activity of non-purified O_1 BFS 1860 harvests, T.R. Doel, unpublished observations) and repeated application does not induce cross-protection (Black et al. 1986). A partial explanation for the data of Cartwright et al. is that low levels of cross-reactivity between serotypes are not infrequent in the serum neutralisation test (Cottral 1972).

FMD vaccines formulated from non-purified antigen harvests certainly contain quantities of NSPs coded for by the viral nucleic acid. These proteins are capable of stimulating specific antibodies in the host and, as indicated above in this chapter, are proving a valuable tool in the development of new assay systems which allow the discrimination between an animal receiving vaccine alone and one which has recovered from infection. A prerequisite for this level of discrimination is the use of vaccines made from purified antigens where the bulk of the NS proteins has been removed. In the author's laboratory, concentrated inactivated FMD antigens are purified by industrial-scale chromatography before a final concentration step and storage over liquid nitrogen until required for formulation. Of course, FMD vaccines are applied repeatedly, and it is essential to demonstrate that a given vaccine does not induce significant titres of antibodies against NSPs under 'worst case scenarios'. We used a procedure in which cattle were vaccinated between three and five times with maximum payload quadrivalent vaccines where each dose was equivalent to approximately 10 to 20 'normal' cattle doses. Under these circumstances, none of the cattle was scored positive as defined by the EITB assay (Bergmann et al. 1993).

References

Abu Elzein EME, Crowther JR (1981) Detection and quantification of IgM, IgA, IgG_1 and IgG_2 antibodies against foot-and-mouth disease virus from bovine sera using an enzyme-linked immunosorbent assay. J. Hyg. Camb 86:79–85

Ahl R, Haas B, Lorenz RJ, Wittmann G (1990) Alternative potency test of FMD vaccines and results of comparative antibody assays in different cell systems and ELISA. In Report of the Session of the Research Group of the Standing Technical Committee of the European Commission for the Control of Foot-and-Mouth Disease. Lindholm, Denmark, 25–29 June. Food and Agriculture Organisation (FAO), Rome, 51–60

Anderson EC, Doughty WJ, Anderson J (1976) The effect of repeated vaccination in an enzootic foot-and-mouth disease area on the incidence of virus carrier cattle. J. Hyg. Camb 73:229–235

Bahnemann HG (1990) Inactivation of viral antigens for vaccine preparation with particular reference to the application of binary ethyleneimine. Vaccine 8:299–303

Barnett PV, Cox SJ, Aggarwal N, Gerber H, McCullough KC (2002) Further studies on the early protective responses of pigs following immunisation with high potency foot and mouth disease vaccine. Vaccine 20:3197–3208

Barteling SJ, Vreeswijk J (1991) Developments in foot-and-mouth disease vaccines. Vaccine 9:75–88.

Bergmann IE, Augé De Mello P, Neitzert E, Beck E, Gomes I (1993) Diagnosis of persistent aphthovirus infection and its differentiation from vaccination responses in cattle by use of enzyme-linked immunoelectrotransfer blot analysis with bioengineered nonstructural viral antigens. Am. J. Vet. Res 54:825–831

Black L, Rweyemamu MM, Boge A (1984) Revaccination response of cattle as a function of the 140S foot-and-mouth disease antigen concentration. J. Comp. Pathol. 94:417–424

Black L, Nicholls MJ, Rweyemamu MM, Ferrari R, Zunino MA (1986) Foot-and-mouth disease vaccination: a multifactorial study of the influence of antigen dose and potentially competitive immunogens on the response of cattle of different ages. Res. Vet. Sci. 40:303–307

Borca MV, Fernández FM, Sadir AM, Braun M, Schudel AA (1986) Immune response to foot-and-mouth disease virus in a murine experimental model: effective thymus-independent primary and secondary reaction. Immunology 59:261–267

Cartwright B, Chapman WG, Sharpe RT (1982) Stimulation by heterotypic antigens of foot-and-mouth disease virus antibodies in vaccinated cattle. Res. Vet. Sci. 32:338–342

Childerstone AJ, Cedillo-Baron L, Foster-Cuevas M, Parkhouse ME (1999) Demonstration of bovine $CD8^+$ T-cell responses to foot-and-mouth disease virus. J. Gen. Virol. 80:663–669

Collen T, Pullen L, Doel TR (1989) T cell-dependent induction of antibody against foot-and-mouth disease virus in a mouse model. J. Gen. Virol. 70:395-403

Collen T, Doel TR (1990) Heterotypic recognition of foot-and-mouth disease virus by cattle lymphocytes. J. Gen. Virol. 71:309–315

Collen T (1991) T cell responses of cattle to foot-and-mouth disease virus. Ph.D. Thesis. Council for National Academic Awards, London

Cottral GE, Gailiunas P (1972) Experimental multiple infection of animals with foot-and-mouth disease viruses. Proc. Am. Mtg. U.S. Anim. Hlth. Assoc. 75:441–465

Cottral GE (1969) Persistence of foot-and-mouth disease virus in animals, their products and the environment. Bull. Off. int. Epiz 71: (3–4) 549–568

Cottral GE (1972) Foot-and-mouth disease virus neutralization test cross reactions. Bull.Off.int.Epiz 77: (7–8) 1239–1261

Cowan KM (1973) Antibody responses to viral antigens. Adv. Immunol. 17:195–255

Cunliffe HR (1962) Antibody response in a group of swine after infection with foot-and-mouth disease virus. Can. J. Comp. Med. Vet. Sci. 26:182–185

Cunliffe HR (1964) Observations on the duration of immunity in cattle after experimental infection with foot-and-mouth disease. Cornell Veterinarian 54:501–510

Dellers RW, Hyde JL (1964) Response of sheep to experimental infection with foot-and-mouth disease virus. Am. J. Vet. Res. 25:469–473

De Leeuw PW, Tiessink JWA, Van Bekkum JG (1979) The challenge of vaccinated pigs with foot and mouth disease virus. Zbl. Vet. Med.B. 26:85–97

Doel TR, Baccarini PJ (1981) Thermal stability of FMDV. Arch. Virol. 70:21-32

Doel TR, Chong WKT (1982) Comparative immunogenicity of 146S, 75S, and 12S particles of FMDV. Arch. Virol. 73:185-191

Doel TR, Collen T (1982) Qualitative assessment of 146S particles of FMDV in preparations destined for vaccines. J. Biol. Stand. 10:69-81

Doel TR, Williams L, Barnett PV (1994) Emergency vaccination against foot-and-mouth disease: rate of development of immunity and its implications for the carrier state. Vaccine 12:592–600

Dubourget PH, Detraz N, Stellmann C, Tixier G, Lombard M (1987) Prophylaxis of foot-and-mouth disease: influence of annual booster vaccination on the level and specificity of neutralizing antibodies. Report of the Meeting of the Research Group of the Standing Technical Committee of the European Commission for the Control of Foot-and-Mouth Disease. Lyon, France. Food and Agriculture Organization of the United Nations

Fish RC, Van Bekkum JG, Lehmann RP, Richardson GV (1969) Immunologic responses in Dutch cattle vaccinated with foot-and-mouth disease vaccines under field conditions: Neutralizing antibody responses to O, A and C types. Am. J. Vet. Res 30:2115–2123

Francis MJ, Black L (1983) Antibody responses in pig nasal fluid and serum following foot-and-mouth disease infection or vaccination. J. Hyg. Camb 91:329–334

Francis MJ, Ouldridge EJ, Black L (1983) Antibody response in bovine pharyngeal fluid following foot-and-mouth disease vaccination and, or, exposure to live virus. Res. Vet. Sci. 35:206–210

Garcia-Valcarcel Munoz-Repiso M (1993) Cellular immune recognition of foot-and-mouth disease virus and derived antigens. Ph.D. Thesis, University of Hertfordshire

Garland AJM (1974) Inhibitory activity of secretions in cattle against foot-and-mouth disease virus. Ph.D. Thesis, University of London

Gibson CF, Donaldson AI, Ferris NP (1984) Response of sheep vaccinated with large doses of vaccine to challenge by airborne foot-and-mouth disease virus. Vaccine 2:157–161

Glass EJ, Oliver RA, Collen T, Doel TR, Dimarchi R, Spooner RL (1991) MHC class II restricted recognition of FMDV peptides by bovine T cells. Immunology 74:594–599

Gomes I (1977) Foot-and-mouth disease: reaction of convalescent pigs to homologous virus exposure. Bol. Centr. Panam. Fiebre Aftosa 26:18–22

Gomes I, Sutmöller P, Casas Olascoaga R (1980) Response of cattle to foot-and-mouth disease (FMD) virus exposure one year after immunization with oil-adjuvanted FMD vaccine. Bol. Centr. Panam. Fiebre Aftosa 37–38:31–35

Graves JH, McKercher PD, Farris HE, Cowan KM (1968) Early response of cattle and swine to inactivated foot-and-mouth disease vaccine. Res.Vet.Sci. 9:35–40.

Hedger RS (1968) The isolation and characterisation of foot-and-mouth disease virus from clinically normal herds of cattle in Botswana. J. Hyg. Camb. 66:27–36

Hendriksen CFM (1988) Laboratory animals in vaccine production and control: replacement, reduction and refinement. Kluwer Academic Publishers: p.131

Kitching RP, Salt JS (1995) The interference by maternally-derived antibody with active immunization of farm animals against foot and mouth disease. Br. Vet. J. 151:379–389

Kitching RP (1997). Vaccination of calves against FMD in the presence of maternally derived antibody. In Report of the Session of the Research Group of the Standing Technical Committee of the European Commission for the Control of Foot-and-Mouth Disease. Kibbutz Ma'ale Hachamisha, Israel, 2–6 September 1996. Food and Agriculture Organisation (FAO), Rome, 191–195

Knudsen RC, Groocock CM, Anderson AA (1979) Immunity to foot-and-mouth disease virus in guinea pigs: clinical and immune responses. Infect. Immun. 24:787–792

Löffler F, Frosch P (1897) Summarischer Bericht über die Ergebnisse der Untersuchungen der Kommission zur Erforschung der Maul- und Klauenseuche bei dem Institut für Infektionskrankheiten in Berlin. Cent. Bakt. Parasit. Infekt. 22: (10/11), 257–259

Manual of Standards for Diagnostic Tests and Vaccines (2000). Foot and Mouth Disease, Chapter 2.1.1. Office International des Epizooties, Paris.

McKercher PD, Giordano AR (1967) Foot-and-mouth disease in swine. I. The immune response of swine to chemically-treated and non-treated foot-and-mouth disease virus. Arch. ges. Virusforsch 20 (1): 39–53

McKercher PD, Graves JH, Cunliffe H, Callis JJ, Fernandez MV, Martins IA, Alonso Fernandez A, Gomes I, Auge De Mello P, Palacios CA (1975a) Foot-and-mouth disease vaccines. I. Comparison of vaccines prepared from virus inactivated with formalin and adsorbed on aluminium hydroxide gel with saponin and virus inactivated with acetylethyleneimine and emulsified with incomplete Freund's adjuvant. Bol. Centr. Panam. Fiebre Aftosa 19–20:9-16

McKercher PD, Graves JH, Cunliffe H, Callis JJ, Auge De Mello P, Gomes I, Alonso Fernandez A, Fernandez MV (1975b) Foot-and-mouth disease vaccines. II. Studies on the duration of immunity in cattle and pigs. Bol. Centr. Panam. Fiebre Aftosa 19–20:24–30

McVicar JW, Sutmöller P (1974) Neutralizing activity in the serum and oesophageal-pharyngeal fluid of cattle after exposure to foot-and-mouth disease virus and subsequent re-exposure. Arch. ges. Virusforsch 44:173–176

Mulcahy G, Gale C, Robertson P, Iyisan S, Dimarchi R, Doel TR (1990) Isotype responses of infected, vaccinated and peptide-vaccinated cattle to FMDV. Vaccine 8:249–256

Newman JFE, Cartwright B, Doel TR, Brown F (1979) Purification and identification of the RNA-dependent RNA polymerase of FMDV. J. Gen. Virol. 45:497-507

Nicholls MJ, Black L, Rweyemamu MM, Gradwell DV (1985) Effect of age on response of cattle to vaccination against foot-and-mouth disease. Br. Vet. J. 141:17–233

Pay TWF (1984) Factors influencing the performance of foot-and-mouth disease vaccines under field conditions. In: Applied Virology. Ed. Kurstak. E. Academic Press 73–86

Pinto AA, Garland AJM (1979) Virus-infection associated (VIA) antigen in cattle repeatedly vaccinated with foot-and-mouth disease virus inactivated by formalin or acetylethyleneimine. J. Hyg. Camb. 82:41–50

Remond M, Cruciere C, Kaiser C, Lebreton F, Moutou F (1998) Preliminary results of a serological survey for residual foot and mouth disease antibodies in French cattle six years after the end of vaccination. In Report of the Session of the Research Group of the Standing Technical Committee of the European Commission for the Control of Foot-and-Mouth Disease. Poiana-Brasov, Romania, 23–27 September 1997, Food and Agriculture Organisation (FAO), Rome, 84–86

Rocha JR, Barrera J, Bustos M (1983) Immune response induced by oil-adjuvanted foot-and-mouth disease vaccine in cattle in tropical areas of Colombia. Bol. Centr. Panam. Fiebre Aftosa 47–48:45–54

Russell WMS (2000). Forty Years On. In 'Progress in the Reduction, Refinement and Replacement of Animal Experimentation'. Eds. M. Balls, A.-M. van Zeller and M.E. Halder. Elsevier, 7–14

Rweyemamu MM, Black L, Boge A, Thorne AC, Terry G (1984) The relationship between the 140S antigen dose in aqueous foot-and-mouth disease vaccines and the serum antibody response of cattle. J. Biol. Stand 12:111–120

Saiz JC, Rodriguez M, Gonzalez M, Alonso F, Sobrino F (1992) Heterotypic lymphoproliferative response in pigs vaccinated with foot-and-mouth disease virus. Involvement of isolated capsid proteins. J. Gen. Virol 73:2601–2607

Salt JS, Barnett PV, Dani P, Williams L (1998) Emergency vaccination of pigs against foot and mouth disease: protection against disease and reduction in contact transmission. Vaccine 16:746–754

Salt JS (1993) The carrier state in foot-and-mouth disease—an immunological review. Br. Vet. J. 149:207–223

Sellers RF, Herniman KAJ (1974) Early protection of pigs against foot-and-mouth disease. Br. Vet. J. 130:440–445

Sharma R, Presad S, Ahuja KL, Rahman MM, Kumar A (1985) Cell mediated immune response following foot-and-mouth disease vaccination in buffalo calves. Acta Virol. 29:509–513

Siskind GW, Benacerraf B (1969) Cell selection by antigen in the immune response. Adv. Immunol. 10:1-50

Sorensen KJ, Hansen CM, Madsen ES, Madsen KG (1998) Blocking Elisas using the FMDV non-structural proteins 3D, 3AB, and 3ABC produced in the baculovirus expression system. Proceedings of the final meeting of concerted action CT93 0909. The Veterinary Quarterly Vol 20: S17

Steward MW, Stanley CM, Dimarchi R, Mulcahy G, Doel TR (1991) High-affinity antibody induced by immunization with a synthetic peptide is associated with protection of cattle against foot-and-mouth disease. Immunology 72:99–103

Sutmöller P (1971) Persistent foot-and-mouth disease virus infections. In 'Viruses affecting man and animals'. Ed. M.Sanders. Warren H. Green Inc, St. Louis, Mo. 295–308B

Sutmöller P, Graves JH, McVicar JW (1970) Influence of enterovirus on foot-and-mouth disease virus infection: a hypothesis. Proc. 74th Ann. Mtg. U.S. Anim. Hlth. Assoc. 235–240

Van Lierop MC, Van Mannen K, Meloen RH, Rutten VPMG, De Jong MAC, Hensen EJ (1992) Proliferative lymphocyte responses to foot-and-mouth disease virus and three FMDV peptides after vaccination or immunization with these peptides in cattle. Immunology 75:406–413

Van Lierop MC, Wagenaar JPA, Noort JM, Hensen EJ (1995) Sequences derived from the highly antigenic VP1 region 140 to 160 of foot and mouth disease virus do not prime for a bovine T-cell response against intact virus. J. Virol. 69:4511–4514

Zhang ZD, Hutching G, Kitching P, Alexandersen S (2002) The effects of gamma interferon on replication of foot-and-mouth disease virus in persistently infected bovine cells. Arch. Virol. 147:2157–2167

Zinkernagel RM, Lamarre A, Ciurea A, Hunziker L, Ochsenbein AF, McCoy KD, Fehr T, Bachmann MF, Kalinke U, Hengartner H (2001) Neutralizing Antiviral Antibody Responses. Adv. Immunol. 1–53

CTMI (2005) 288:133–148

Global Epidemiology and Prospects for Control of Foot-and-Mouth Disease

R. P. Kitching

National Centre for Foreign Animal Disease, 1015 Arlington Street, Winnipeg, Manitoba, R3E 3M4, Canada
rpkit@lineone.net

1 Introduction

Despite the dramatic advances in our understanding of viral pathogenesis and the development of vaccine technology brought about by the enlightenment of the most intimate secrets of how viruses interact with their host cells, foot-and-mouth disease (FMD) virus remains a major threat to the most sophisticated economies of the world. The combined threats of free trade and bio-terrorism have shown how vulnerable the agricultural industries of North America and Europe are to attack by one of the smallest living organisms; an attack against which the defences are little better than they were 50 years ago. It has been proposed in the past that the best protection for the developed countries that are free of FMD would be to eradicate the virus from countries in which it is endemic, but the events of the last few years clearly show that FMD is in the ascendency, and is far from dead. The Food and Agriculture Organization (FAO) is to make FMD its next target for global eradication, following its not yet fully successful eradication of rinderpest. But compared with rinderpest that is caused by a single virus serotype, against which there is a very effective vaccine that provides virtual lifelong immunity after a single inoculation, that does not produce persistent infection and that has a very limited host range, any program to eradicate FMD will certainly fail. The tools are not yet available for such a task. The best that can be expected with current resources would be an attempt to bring the disease under control in as many of the endemic countries of the world as can sustain the recurrent cost of vaccine and the imposition of rigid animal movement restrictions. This is an unlikely proposition in those countries which see little economic benefit for such cost, and therefore, until the consequences of the epidemiological differences between rinderpest and FMD are reduced by the development of a better vaccine or a cheap virucidal drug, FMD will remain a significant threat for the foreseeable future.

2
Epidemiology of Foot-and-Mouth Disease

It is unwise to consider FMD as a single disease which always behaves in a pre-determined manner. To do so can be an economically and socially expensive mistake, as was discovered by the British government when it chose to follow the advice of the modellers to bring the 2001 outbreak of FMD under control by slaughtering large numbers of animals which, because of the particular nature of the Pan Asia strain causing the outbreak, were never likely to have been exposed. The models used at the start of the outbreak made no allowance for the differences between this strain and those on which they had been based, even though it was evident from the Pan Asia outbreaks in Japan and South Korea the previous year, and from the limited spread from the index case in the UK, that this virus was epidemiologically distinct.

Foot-and-mouth disease is seven separate diseases, clinically indistinguishable, caused by seven antigenically distinct serotypes. And even within each serotype there is a spectrum of strains with their own antigenic and epidemiological characteristics, which make it impossible to generalise about what to expect in an outbreak (Kitching 1998). The pig-specific strains of serotype O FMD virus found in south-east Asia behave very differently from those found in South America. The epidemiology of the SAT serotypes differ from each other, although all remain restricted to Africa, and when they do escape into the Middle East, they never persist. The ephemeral behaviour of serotype C, which in the last few years has only been seen in East Africa, is difficult to explain. There is a constant emergence of antigenically novel strains of serotype A, both in Asia and South America, but they tend to disappear just as regularly. New strains of serotype ASIA1 can also be found, but antigenically the group remains relatively stable.

3
Transmission of Foot-and-Mouth Disease Virus

Although strains of FMD virus exhibit different epidemiologies, they are constrained by certain common rules, many of which are also shared by other viruses. Spread of FMD virus is most commonly associated with the movement of infected animals and their contact with susceptible animals. During the early stages of disease, infected animals shed virus in all their excretions and secretions, including their breath. In cattle and pigs peak production of virus coincides with the onset of clinical signs, whereas in sheep it occurs before the appearance of lesions, and then in all species it declines rapidly as antibody production and other immune responses bring the infection under control. Virus may infect in-contact animals by the oral or aerosol route, or through skin abrasions, particularly in pigs. The incubation period for FMD is between 1 and 14 days, depending on the strain of virus, the infecting dose, the route of infection and the susceptibility of the host. The minimum infecting doses for some strains have been determined (Sellers 1971; Donaldson and Alexandersen 2002; Sutmoller and Vose 1997).

Clinical FMD in North American and European breeds of cattle and sheep which have not received vaccination is usually not difficult to identify, particularly once it has become established on a farm. Spread between cattle and between pigs is usually rapid so that frequently 90% of the animals may be showing signs. Sheep and goats show milder clinical signs of FMD, and it is not uncommon to confuse them with other conditions that cause foot or mouth lesions (De la Rua et al. 2001). Spread within an infected sheep flock is also slower, and it was evident from the outbreaks of FMD in Italy in 1993 and Greece in 1994 and 1996 that the virus disappeared from many of the flocks before all the animals had been infected. In some flocks only 25% of the animals had been infected before the virus had died out. Kitching and Hughes (2002) have described the survival of FMD virus in groups of sheep under experimental conditions, and their results confirm the field observations that transmission of FMD virus between sheep can be precarious. However, under the conditions of crowding and frequent movement which occurred in the UK during the 2001 outbreak, sheep played a crucial role in spreading the disease around the country and into Ireland and France. Some breeds of cattle also show a degree of natural resistance to clinical FMD, for instance the Brahmin cattle of sub-Saharan Africa, and the Chinese Yellow breed of cattle (Kitching 2002).

FMD virus undoubtedly entered the UK in 2001 in illegally imported meat products, probably from south-east Asia—possibly by the same route by which hog cholera virus had entered the UK the year before. Some of this meat was infected with a strain of the Pan Asia serotype O topotype, which was prevalent in Asia, and some eventually was fed to pigs (see below). All products from animals slaughtered with clinical FMD or while incubating the disease will contain the virus; however, by allowing a slaughtered carcase to hang at 2°C for 24 h before freezing, the change in pH in the muscle to below 6 is sufficient to kill the virus. It will still be necessary to cut the meat from the bone and separate the glands, as the same pH change will not occur in these organs. Milk and semen can be contaminated with FMD virus for up to 4 days before the appearance of clinical signs, but, assuming only a few animals are infected before the disease is recognized and milk collection stopped, by the time the infected milk has been diluted and pasteurized, it will be unlikely to contain an infectious dose (Donaldson 1997). Milk tankers carrying infected milk onto an uninfected farm can spread disease by venting their tank as they take on additional milk; this occurred during the 1967/68 outbreak in the UK and also during the most recent outbreak (Gibbens et al. 2001). To spread disease the infected product must have contact with an FMD-susceptible animal, and therefore typically it is pigs which are the first infected in an outbreak of FMD in a previously free country.

FMD virus can also be carried mechanically by people, vehicles, brushes, surgical equipment and other fomites from infected to susceptible animals. Veterinarians were involved in the spread of FMD in Denmark in 1982, and in Italy in 1993—the former using contaminated surgical equipment and the latter carrying out artificial insemination after being on an infected farm. Similarly, during the 2001 outbreak, farmers were implicated in the spread of virus between sheep flocks. But, as with infected animal products, fairly intimate contact must be made between the contaminated fomite and the susceptible animal for infection to occur.

During the early acute phase of FMD, infected animals excrete virus as they breathe. The quantity depends on the host species and the virus strain, and early work by Donaldson et al. (1982) identified the pig as the largest producer of aerosol virus. Log10 8.6 $TCID_{50}$ can be produced by a pig in a 24-h period when infected with a strain of serotype C. Cattle and sheep produce about 3,000 times less, and therefore they are considerably less significant than pigs as a source of aerosol virus as a means of transmitting disease between farms (Donaldson et al. 2001). There is a large variation between strains in the amount of aerosol virus produced from infected animals; for example, Alexandersen and Donaldson (2002) measured only Log10 6.1 $TCID_{50}$ of virus from pigs infected with the PanAsia strain that infected the UK, 300 times less than that produced by pigs infected with the C Noville strain—consistent with the field observations (see above).

When produced in sufficient quantity, and under ideal weather conditions of relative humidity above 60%, light breeze in one direction, flat topography and climatic inversion keeping the virus plume close to the ground, FMD virus can spread as an aerosol a considerable distance. In 1981, FMD spread from Brittany in France to the Isle of Wight in southern England, a distance of over 250 km., and there are other examples of aerosol spread over large distances (Donaldson and Alexandersen 2002). It is usually cattle which become infected by aerosol virus because of their extreme susceptibility to this route of infection, and their large respiratory tidal volume compared with smaller ruminants, which are also very susceptible to aerosol virus. Cattle may become infected at concentrations of FMD virus as low as 0.06 $TCID_{50}$ per cubic metre of air (Donaldson et al. 2001). Pigs are considerably less susceptible to infection by aerosol virus, and may require up to 6,000 times higher concentrations (Donaldson and Alexandersen 2001).

Ruminants, in particular cattle and the African buffalo, will retain live FMD virus in the cells of the pharyngeal epithelium for up to 3 years in cattle, and longer in buffalo, after initial infection. The immune state of the bovine at the time of contact with the live virus, whether protected from disease by vaccination or fully susceptible, does not affect the establishment of the carrier state, and over 50% of cattle will become carriers—as defined by the recovery of live virus 28 days or longer after infection. The potential for these carrier animals to cause fresh outbreaks of FMD is extremely controversial and not proven (Kitching 2002), but the possibility that a bovine vaccinated during an outbreak of FMD could be carrying live virus has a profound influence on the attitude of the international community to the use of FMD vaccine. The Netherlands slaughtered all 200,000 cattle that had been vaccinated during the 2001 outbreak in order to re-establish its trading status as quickly as possible.

Under experimental conditions it has never been possible to show transmission of FMD virus from a carrier to a susceptible bovine. This does not establish that it cannot happen under certain circumstances, and field situations in which it appears that outbreaks have been caused by carriers invariably involve the movement and mixing of the carriers and otherwise causing them "stress". The relative distributions of the different serotypes and even topotypes of FMD virus, and their restriction and persistence in certain geographical areas, does suggest that some strains adapt to certain hosts, and the existence of the carrier state could provide at least a partial explanation.

4 Serotype O

FMD caused by serotype O has always been the most dominant and most widely distributed. It has the reputation of being the most aggressive serotype and the most difficult to control by vaccination, although so far there is no genetic explanation as to why this serotype is so invasive. Some of the strains are clearly catholic in their host range, appearing equally virulent and transmissible in sheep, goats, pigs and cattle, whereas others, like those of the Cathay topotype, will only cause disease in pigs, even under experimental conditions (Dunn and Donaldson 1997). The Cathay topotype is found in the densely pig-populated countries of south-east Asia and appears to be maintained by the feeding to susceptible pigs of waste food contaminated with virus. Spread of the virus between pigs is by close contact, and investigation of outbreaks in the Philippines, where it is now restricted to "backyard" pigs, has shown that the virus does not transmit even between neighbouring backyards as an aerosol. In Hong Kong, where this topotype is also common, the abattoir lairage became contaminated with virus, and pigs waiting for slaughter became infected and were developing lesions within 24 h of entry. These pigs were then slaughtered while viraemic, with the consequence that all their meat was contaminated with virus, and therefore, following the not uncommon practice of feeding waste restaurant and other waste food to growing pigs, the disease was perpetuated. It is currently the practice in Hong Kong to vaccinate pigs going for slaughter in order to give them some protection against FMD while waiting in the lairage.

Not that the feeding of waste food to pigs is restricted to Asia. The index case for the UK 2001 outbreak was on a farm in Northumberland which fed waste food from restaurants and other sources (swill). Countries that are free of FMD follow the OIE guidelines which prevent the importation of FMD virus in live animals and their products from countries in which the virus is present. However, it is recognized that illegal importation of meat is likely to occur, and additional safeguards are in place whereby all swill must be boiled to kill any contaminating virus (not only FMD virus, but also hog cholera virus, swine vesicular disease virus and others) before feeding to pigs. The owner of this index farm has since been prosecuted for not boiling the swill, and swill feeding throughout the EU has now been banned. The UK outbreak was caused by a strain of the PanAsia topotype, and like the Cathay topotype, its epidemiology was not characterised by significant aerosol spread. Undoubtedly, aerosol transmission did occur to some of the neighbouring farms, but considering that there were almost 500 clinically infected adult pigs on the farm, and that the virus had been present for at least 3 weeks before they were slaughtered, if this virus had been produced as an aerosol in the same quantity as had been shown for some other strains (see above), infection would have occurred in cattle on the coast of Denmark.

Unlike the Cathay topotype, the PanAsia topotype is not restricted to infecting pigs. In the UK it caused clinical disease in cattle, pigs, sheep and goats, although its effect on sheep was less obvious (see above). In South Korea during the 2000 outbreak it affected only cattle, whereas in the 2002 outbreak it affected only pigs. When, in 1999, it was discovered in Taiwan, a strain of this topotype failed to produce clinical disease in Chinese Yellow cattle but caused outbreaks in cattle and goats. In Mongolia, a PanAsia strain reportedly was also killing Asian camels during their 2000 outbreak. The PanAsia topotype has been extremely successful in gaining entry into countries previously free of FMD—Japan had been free since 1908, South Korea since 1934 and the UK free since 1981. It has also out-competed other topotypes of serotype O in the Middle East and is now the predominant group identified in outbreaks. It also caused an outbreak in 2000 in South Africa, having gained entry in untreated swill collected from a boat in Durban harbour and fed to pigs, and South Africa had never before recorded an outbreak of FMD caused by serotype O. What epidemiological characteristic of this topotype has allowed it to be so successful is not clear, certainly not long-distance aerosol transmission or antigenic variability.

Serotype O has in 2001 been active in South America, causing outbreaks in Argentina, Uruguay and southern Brazil, all areas that had only recently stopped prophylactic vaccination. In 2002, Paraguay reported the presence of serotype O in vaccinated cattle close to the Brazilian border. The South American strains of serotype O are characterised by aerosol spread, and it was after the introduction of one of these strains from Argentina into the UK in 1967 that it was realized that between-farm aerosol transmission had been occurring (Hugh-Jones and Wright 1970). It is more likely, however, that in the South American situation, illegal movement of infected animals would account for most of the spread that was seen.

5
Serotype A

Serotypes were defined by the lack of immunity in cattle provided by infection with one serotype against challenge with another. Using this definition it would be possible to split strains of serotype A into at least two new serotypes, as there is no cross-immunity between, for instance, A22 Iraq and A24 Cruzeiro—although strains are no longer placed in subtypes (see Kitching et al. 1989), old subtype designations are still used for historic strains of FMD virus. There is greater antigenic diversity between serotype A strains than between strains within serotypes O, C or ASIA 1, and there has been a constant and frequent appearance of antigenically novel strains, particularly in western Asia. The World Reference Laboratory for FMD maintains a database of more than 2,000 FMD virus partial 1D gene sequences, which can be rapidly compared with the sequence of a new outbreak strain to identify a possible origin. In 1996, outbreaks due to a strain of serotype A were causing major concern to the dairy industry in Iran, because of the lack of immunity provided by the current vaccine. When the region of the IP4 gene was sequenced, it was 18% different from anything in the database, which was very unusual, as previously even a 10% difference from existing strains, other than within the SAT serotypes, was rare (Kitching 1998). It is possible to speculate that the virus had been circulating in an area from which samples were rarely if ever sent to the World Reference Laboratory, that it had been maintained in a species, such as the Asian buffalo, from which samples are seldom collected, or that a large and fairly rapid mutation event had occurred. In 1999, another antigenically new strain of serotype A appeared in Iran, although its distant antecedents could be identified. Antigenically new strains have also appeared in Brazil in 1993, in Columbia in 1985 and in Eritrea in 1998, and other examples are frequent. The significance of these new strains is that they make control by vaccination very difficult, both for the producer who suffers the consequence of an outbreak and for the vaccine producer who must invest in the development of another vaccine strain. But, paradoxically, many of these new strains disappear just as quickly, without necessarily creating the huge outbreaks in the unprotected population as was predicted.

6 Serotype C

During the last 20 years there have been no major outbreaks reported due to strains of serotype C. Sporadic outbreaks have occurred from which a strain of serotype C was isolated, particularly in Nepal, Bhutan and India. More recently, only Kenya has been reporting outbreaks due to serotype C, and these may be associated with the continuing use of a strain of serotype C in the locally produced vaccine. There is no obvious explanation to account for the apparent disappearance of serotype C from the world, although it may still be present.

7 Serotype ASIA 1

Strains of ASIA 1 serotype anecdotally appear less aggressive than those of other serotypes in terms of a longer incubation period and better vaccinal protection. The serotype is restricted to Asia, although for the first time it caused an outbreak in Greece in 2000 on the border with Turkey. It is not clear why ASIA 1 has never been reported in Africa, as there must have been many opportunities for strains to spread to Africa in infected animals. In the Middle East it is not unusual to isolate a strain of ASIA 1 together with a strain from another serotype from the same animal, and Woodbury et al.(1994) were able to show that strains from different serotypes could transmit together. On one occasion strains from serotypes O, A and ASIA 1 were isolated from a single bovine sample collected in Saudi Arabia, which would indicate that antigenically divergent strains from the same serotype could also co-exist in the same host.

8 SAT Serotypes

The SAT serotypes, like ASIA 1, are geographically restricted to Africa. The major movement of animals is from Africa to Asia, and little attempt is made to prevent the export of FMD-infected animals. There have been a number of occasions when SAT strains have been found in the Middle East, the most recent being SAT 2 in Saudi Arabia in 2000. However, although there were large outbreaks in the unprotected dairy herds—SAT strains were not included in the routine vaccination program—the virus does not appear to have persisted in the region. This is certainly not because of a successful vaccination campaign, as the only effective vaccination is practised by the large dairy herds and must be related to a host preference, or even the ability to establish the carrier state in the local animal population. During and after the SAT 2 outbreaks in Zimbabwe in 1991, oro-pharyngeal (probang) sampling of the local Brahmin and imported Friesian cattle indicated that the virus persisted considerably longer in the Brahmin—one Brahmin bull carried for over 3-1/2 years, whereas no Friesian cattle were identified as carriers after 12 months (Dawes, personal communication).

SAT 2 is the more common serotype isolated from cattle in Africa. SAT 1 and 3 are maintained in the African buffalo, occasionally causing outbreaks in cattle with which the buffalo have contact or in impala, which then spread disease to the cattle (Bastos et al. 2000). The African buffalo can remain a carrier of SAT viruses for over 5 years, longer than that recorded for cattle, particularly with these serotypes, which adds support to the hypothesis of the importance of the carrier state for the survival of the FMD virus.

There is considerable sequence variation between strains within each of the SAT serotypes, suggesting that there could be considerable antigenic variation (Bastos et al. 2001; Vosloo et al. 1995, 1996). The economic rewards of investigating these differences and producing separate vaccine strains for each antigenically distinct group within each serotype would not be justified by any of the vaccine producers, bearing in mind the limited use of FMD vaccine within the SAT endemic countries.

9
Prospects for Control

It is self-evident that the control of FMD depends on controlling its spread from an infected to a susceptible animal, either by preventing the movement of the virus in infected animals, animal products, fomites, aerosol or carriers or by reducing the number of susceptibles by vaccination. Countries usually free of FMD usually attempt the former by strict import controls, as was seen in Europe in 2001, not always successfully. However, the benefit of maintaining the national flock and herd free of FMD without the use of vaccination, in terms of increased export markets and without the regular expense of vaccine, was considered to exceed the cost of eliminating the occasional outbreak of disease and encouraged the EU to stop prophylactic vaccination at the end of 1991. The cost of the UK outbreak, which has been estimated at 12 billion dollars US, has caused a re-evaluation, not only of the financial cost, but also of the social and welfare cost of controlling and eliminating an outbreak by slaughter. That at least 2 million animals were slaughtered unnecessarily is probably now irrelevant; the image remains of huge piles of corpses waiting disposal. At the International Conference on Control of FMD held in Brussels in December 2001, it was made quite clear by the EU politicians that slaughter on the scale that occurred during the 2001 outbreaks in Europe was no longer politically acceptable and that an alternative was required, in particular, the early use of vaccination, as there was now a serological test to distinguish animals that were antibody positive from vaccination from those that were positive after infection.

10
Vaccination

Although the compulsory use of vaccine against FMD on most of mainland Europe in the 1960s coincided with a dramatic drop in the number of outbreaks, this was not a consequence of vaccination alone. At the same time, outbreaks were being controlled by slaughter of infected and in-contact susceptible animals, disinfection to kill the virus on infected farms, movement controls around infected farms to prevent movement of infected livestock and control of the movement of potentially infected animal products. It has long been clear that vaccination using currently available vaccines is not sufficient to prevent FMD—vaccination of ruminants may stop disease but will not stop infection. In some of the large dairy herds of the Arabian peninsula, cattle are vaccinated every 10 weeks with vaccine that contains seven strains of FMD virus, two serotype O, three serotype A, one serotype ASIA 1 and one serotype SAT 2, and yet they still succumb to clinical FMD. Only by using strict security to prevent infected material entering the farm, and maintaining a separation of at least 100 m between the cattle and the perimeter fence (and the livestock belonging to the nomadic people), can the number of outbreaks of FMD be reduced. Vaccination is being used as a second line of defence (Kitching and Hutber 2003). In pigs, vaccination will protect against moderate levels of challenge, but once FMD is present in the herd, the increased amount of virus in the environment will overcome vaccine immunity; this is recognized when carrying out vaccine potency tests in pigs: The unvaccinated control pigs are kept in a separate isolation cubicle, out of contact with the vaccinated animals as, if kept together, when the controls develop FMD, they would likely precipitate disease in the vaccinates.

Apart from not preventing infection or even disease, FMD vaccines provide only short-term immunity, and it is recommended that cattle be re-vaccinated every 4 or 6 months, depending on the expected challenge. Protection is highest against the strain from which the vaccine was made (homologous protection) and becomes progressively lower the more antigenically different the outbreak strain is from the vaccine strain. This relationship can be expressed as an "r" value, and depending on the value of this ratio, it is possible to anticipate the level and duration of protection provided by the vaccine (Kitching et al. 1989). To produce conventional FMD vaccine it is necessary to grow large amounts of virus in tissue culture, under disease-secure conditions, completely inactivate and then purify and concentrate the antigen before mixing with an adjuvant (Doel 1996). These procedures themselves are not without danger of escape of virus, either directly from the facility or because the virus was not properly inactivated before being made into vaccine (Beck and Strohmaier 1987).

It was initially expected that new developments in biotechnology would quickly produce a more effective FMD vaccine, but the early promise soon evaporated (Kitching 1992), as did most of the commercial funding for such projects. Grubman and Mason (2002) have reviewed attempts to produce a better FMD vaccine and describe their own candidate, a replication-defective human adenovirus within which are the 4 capsid and 3C proteinase coding genes of FMD virus. The expressed proteins form empty FMD virus capsids which stimulate protective immunity in pigs and cattle. Another prospect is a peptide vaccine which utilises only a short, but antigenically important, sequence from the 1D gene, together with B and T-helper cell epitopes (Wang et al. 2001). This reportedly protects pigs but has not yet been proven in cattle. Both these potential vaccines have the advantage of not requiring the production of large quantities of live FMD virus, and animals vaccinated with either could easily be distinguished from animals that had recovered from infection by a non-structural protein (NSP) antibody test. Further testing will be required to show whether they prevent the development of the carrier state, the degree of cross-immunity provided between strains and the duration of immunity. It would be expected that the cost of their production would be less than that of conventional vaccine, thus encouraging their use in some of the poorer countries.

In the past, live attenuated FMD vaccines have been used, particularly in South America and China, but they became discredited because of their tendency to revert to virulence. Animals that have recovered from infection remain protected for years rather than the months following vaccination with a dead vaccine, and it is likely that an attenuated vaccine could afford a similar long-term protection. With the advances provided by the technology of reverse genetics, it would be possible to engineer a virus that could not revert to virulence, once, of course, the host specific and virulence determinants of the virus were properly understood.

Considerable enthusiasm has been generated by the development of the NSP antibody tests as a means whereby vaccination can be used to control an outbreak. It is possible to distinguish those animals which have been vaccinated from those which have been infected, by the presence of antibodies to the NSP's in animals that have supported a live replicating FMD virus. Animals receiving a dead virus vaccine have antibodies only to the structural proteins. These tests have not been fully validated, particularly in vaccinated animals which have had contact with live virus and have become carriers. Because the animals will suppress viral replication, there may be insufficient expression of the NSPs to produce a detectable antibody response. Currently such animals represent a very small, but significant, risk. However, on the understanding that the new NSP tests would identify infection, at least on a herd basis, the OIE was persuaded to reduce the time from 12 months to 6 months at which a country, usually free of FMD but which had used vaccine to help control an outbreak, could re-apply for FMD-free status if it had not subsequently slaughtered the vaccinated animals (OIE, 2003). This was conditional on testing a sample of the vaccinated animals for NSP antibodies. It is perhaps not surprising that the EU in its new draft proposal on Community measures for the control of FMD prohibits the movement of vaccinated animals, even after the sero-survey using the NSP tests has shown the vaccinated zone to be infection free to OIE standards (EU, 2002).

11 Conclusion

At present, control of FMD is firmly dependent on the use of vaccine. But current vaccines are expensive and strain specific and provide only short-term immunity. In addition, vaccination alone will not control FMD, and although the additional measures required can be implemented in developed countries, this is not the situation in many of the FMD-endemic countries. Totally protecting FMD-free countries from the intentional or accidental importation of virus would paralyse international trade and travel and would be unacceptable; however, the cost of even a small outbreak in the USA would be considerable (Paarlberg et al. 2002).

An alternative is to eliminate the virus from the world as proposed by FAO, but this would be impossible with current technology. Rweyemamu and Astudillo (2002) have put forward an alternative proposal based on a four-stage pathway, similar to that used for rinderpest control, but rinderpest is as different from FMD as smallpox is from influenza, and in my opinion we still need to understand more about the epidemiology of FMD if such a programme is to be successful.

References

Alexandersen S and Donaldson AI (2002). Further studies to quantify the dose of natural aerosols of foot-and-mouth disease virus for pigs. Epidemiol. Infect., 128, 313–323

Bastos ADS, Boshoff CI, Keet DF, Bengis RG and Thomson GR (2000). Natural transmission of foot-and-mouth disease virus between African buffalo (*Syncerus caffer*) and impala (*Aepyceros melampus*) in the Kruger National Park, South Africa. Epidemiol. Infect., 124, 591–598

Bastos ADS, Haydon DT, Forsberg R et al. (2001). Genetic heterogeneity of SAT-1 type foot-and-mouth disease viruses in southern Africa. Arch. Virol., 146, 1537–1551

Beck E and Strohmaier K (1987). Subtyping of European foot-and-mouth disease virus strains by nucleotide sequence determination. J. Virol., 61, 1621–1629

De la Rua R, Watkins GH and Watson PJ (2001). Idiopathic mouth ulcers in sheep (letter). Vet. Rec., 149, 30–31

Doel TR (1996). Natural and vaccine-induced immunity to foot and mouth disease: the prospects for improved vaccines. Rev. sci. tech. Off. int. Epiz., 15, 883–911

Donaldson AI (1997). Risks of spreading foot and mouth disease through milk and dairy products. Rev. sci. tech. Off. int. Epiz., 16, 117–124

Donaldson AI and Alexandersen S (2001). The relative resistance of pigs to infection by natural aerosols of foot-and-mouth disease virus. Vet. Rec., 148, 600–602

Donaldson AI and Alexandersen S (2002). Predicting the spread of foot and mouth disease by airborne virus. Rev. sci. tech. Off. int. Epiz., 21, 569–575

Donaldson AI, Ferris NP and Gloster J (1982). Air sampling of pigs infected with foot and mouth disease virus : comparison of Litton and cyclone samplers. Res. Vet. Sci., 33, 384–385

Donaldson AI, Alexandersen S, Sorensen JH and Mikkelsen T (2001). Relative risks of uncontrollable (airborne) spread of foot-and-mouth disease by different species. Vet. Rec., 148, 602–604

Dunn CS and Donaldson AI (1997). Natural adaptation to pigs of a Taiwanese isolate of foot-and-mouth disease virus. Vet. Rec., 141, 174–175

EU (2002). Draft proposal for a Council Directive on Community measures for the control of foot-and-mouth disease and amending Directive 92/46/EEC, Article 64, page 59

Gibbens JC, Sharpe CE, Wilesmith JW et al. (2001). Descriptive epidemiology of the 2001 foot and mouth disease epidemic in Great Britain: the first five months. Vet. Rec., 149, 729–743

Grubman MJ and Mason PW (2002). Prospects, including time-frames, for improved foot and mouth disease vaccines. Rev. sci. tech. Off. int. Epiz., 21, 589–600

Hugh-Jones ME and Wright PB (1970). Studies on the 1967–8 foot and mouth disease epidemic, the relation of weather to the spread of disease. J. Hyg., Camb., 68, 253–271

Kitching RP (1992). The application of biotechnology to the control of foot-and-mouth disease virus. Br. Vet. J., 148, 375–388

Kitching RP (1998). A recent history of foot-and-mouth disease. J. Comp. Pathol., 118, 89–108

Kitching RP (2002). Identification of foot and mouth disease virus carrier and subclinically infected animals and differentiation from vaccinated animals. Rev. sci. tech. Off. int. Epiz., 21, 531–538

Kitching RP and Hughes GJ (2002). Clinical variation in foot and mouth disease: sheep and goats. Rev. sci. tech. Off. int. Epiz., 21, 505–512

Kitching RP and Hutber M (2003). The epidemiology of foot and mouth disease in large dairy herds in the Arabian Peninsula. In: Foot and Mouth Disease, Control Strategies. Pub: Merieux Foundation, IAB and OIE (in press)

Kitching RP, Knowles NJ, Samuel AR and Donaldson AI (1989). Development of foot-and-mouth disease strain characterisation—a review. Trop. Anim. Hlth. Prod., 21, 153–166

OIE (2003). International Animal Health Code, Ch. 2.1.1. Foot and Mouth Disease. 13th Edition, OIE, 12, Rue de Prony, Paris (in press)

Paarlberg PL, Lee JG and Seitzinger AH (2002). Potential revenue impact of an outbreak of foot-and-mouth disease in the United States. JAVMA, 220, 988–992

Rweyemamu MM and Astudillo VM (2002). Global perspective for foot and mouth disease control. Rev. sci. tech. Off. int. Epiz., 21, 765–773

Sellers RF (1971). Quantitative aspects of the spread of foot and mouth disease. Vet. Bull., 41, 431–439

Sutmoller P and Vose DJ (1997). Contamination of animal products: the minimum pathogen dose required to initiate infection. Rev. sci. tech. Off. int. Epiz., 16, 30–32

Vosloo W, Kirkbride E, Bengis RG, Keet DF and Thomson GR (1995). Genome variation in the SAT types of foot-and-mouth disease viruses prevalent in buffalo (*Syncerus caffer*) in the Kruger National Park and other regions of southern Africa, 1986–1993. Epidemiol. Infect., 114, 203–218

Vosloo W, Bastos AD, Kirkbride E et al. (1996). Persistent infection of African buffalo (*Syncerus caffer*) with SAT-type foot-and-mouth disease viruses: rate of fixation of mutations, antigenic change and interspecies transmission. J. Gen. Virol., 77, 1457–1467

Wang CY, Chang TY, Walfield AM et al. (2001). Synthetic peptide-based vaccine and diagnostic system for effective control of FMD. Biologicals, 29, 221–228

Woodbury EL, Samuel AR, Knowles NJ, Hafez SM and Kitching RP (1994). Analysis of mixed foot-and-mouth disease virus infections in Saudi Arabia: prolonged circulation of an exotic serotype. Epidemiol. Infect., 112, 201–211

CTMI (2005) 288:149–173

Foot-and-Mouth Disease Virus Evolution: Exploring Pathways Towards Virus Extinction

E. Domingo[1] · N. Pariente[1] · A. Airaksinen[1] · C. González-Lopez[1] · S. Sierra[1] · M. Herrera[1] · A. Grande-Pérez[1] · P. R. Lowenstein[2] · S. C. Manrubia[3] · E. Lázaro[3] · C. Escarmís[1] (✉)

[1] Centro de Biología Molecular "Severo Ochoa" (CSIC-UAM), Universidad Autónoma de Madrid, Cantoblanco, 28049 Madrid, Spain
cescarmis@cbm.uam.es
[2] Gene Therapeutics Research Institute, Cedars-Sinai Medical Center and Department of Medicine, David Geffler School of Medicine, UCLA, Los Angeles, CA 90048, USA
[3] Centro de Astrobiología (CSIC-INTA), Torrejón de Ardoz 28850, Madrid, Spain

Abstract Foot-and-mouth disease virus (FMDV) is genetically and phenotypically variable. As a typical RNA virus, FMDV follows a quasispecies dynamics, with the many biological implications of such a dynamics. Mutant spectra provide a reservoir of FMDV variants, and minority subpopulations may become dominant in response to environmental demands or as a result of statistical fluctuations in population size. Accumulation of mutations in the FMDV genome occurs upon subjecting viral populations to repeated bottleneck events and upon viral replication in the presence of mutagenic base or nucleoside analogs. During serial bottleneck passages, FMDV survive during extended rounds of replication maintaining low average relative fitness, despite linear accumulation of mutations in the consensus genomic sequence. The critical event is the occurrence of a low frequency of compensatory mutations. In contrast, upon replication in the presence of mutagens, the complexity of mutant spectra increases, apparently no compensatory mutations can express their fitness-enhancing potential, and the virus can cross an error threshold for maintenance of genetic information, resulting in virus extinction. Low relative fitness and low viral load favor FMDV extinction in cell culture. The comparison of the molecular basis of resistance to extinction upon bottleneck passage and extinction by enhanced mutagenesis is providing new insights in the understanding of quasispecies dynamics. Such a comparison is contributing to the development of new antiviral strategies based on the transition of viral replication into error catastrophe.

1
Introduction: Concepts in RNA Virus Evolution

Foot-and-mouth disease virus (FMDV) is one of the prototypes of antigenically variable virus, reflected in seven serotypes (A, O, C, Asia 1, SAT1, SAT2, SAT3), and many subtypes and variants, too numerous to be amenable to any reasonable cataloguing at present. The diversity of antigenic types creates difficulties for prevention of FMD by vaccination because there is no predictable, reproducible and effective protection that can be afforded by a limited number of vaccine antigens against multiple variants cocirculating in different world areas, and sometimes even within the same geographical area (Doel 2003; Sutmoller et al. 2003). Traditionally, vaccine manufacturers have known that it is necessary to tailor vaccine composition to match the antigenic properties of the circulating FMDV in much the same way as the influenza vaccines must be periodically updated. Chemically defined vaccines would be desirable, but they will often fail to provide protection against the diverse and heterogeneous array of virus forms to be controlled. Vaccine inefficacy is a major challenge for prevention against diseases caused by variable RNA viruses such as AIDS, hepatitis C and possibly SARS, among many others.

Phenotypic variation of FMDV has been known since the early work of C.R. Pringle, H. Bachrach and others (reviewed in Bachrach 1968 and Domingo et al. 2003). There is now little doubt that antigenic variation of FMDV is a consequence of the error-prone replication of the viral genome, visualized as variability of antigenic epitopes. This feature is shared with other RNA viruses and it is due to the lack (or low efficiency) of proofreading repair and postreplicative repair activities during RNA-dependent RNA and DNA synthesis (Holland et al. 1982; Drake and Holland 1999; Domingo et al. 2001). Mutation rates during RNA replication and retrotranscription average 10^{-3}–10^{-5} misincorporations per nucleotide copied. Together with homologous and nonhomologous recombination and genome segment reassortment (in viruses with segmented genomes), these mechanisms provide the molecular scenario on which virus diversification and adaptability are built (overviews are given by Morse 1993, 1994; Gibbs et al. 1995; Domingo et al. 1999, 2001; Domingo 2003). Despite high error rates affecting all RNA viruses examined to date (Drake and Holland 1999) the extent of antigenic variation, and therefore the resulting problems of low vaccine efficacy derived from variation, do not affect all RNA viruses equally. For example, there is one serotype of Mengo virus, three of poliovirus, seven of FMDV and more than one hundred of human rhinoviruses, despite monoclonal-escape mutants arising with comparable frequencies in the range of 10^{-3}–10^{-5} for these picornaviruses (values have been compared in Domingo et al. 2002). This indicates the existence of constraints to antigenic diversification operating with different intensity, even among viruses sharing a pattern of genetic organization and features of their virion structure. Defining the molecular basis of such constraints remains a challenge for virology, which requires bridging biochemistry with evolutionary biology.

FMDV being a picornavirus of about 8,300 nucleotides with a highly compact genetic information, a single open reading frame, *cis*-acting functions, regulatory regions dependent on precise spatial structures and several multifunctional proteins and from all evidence replicating close to the error threshold for maintenance of genetic information (Sect. 2), it is remarkable that a large repertoire of many different mutations are tolerated by the virus. Mutations may affect important functions such as translation or replication efficiency, cell tropism or host range, and they have been produced during virus replication, without the need of chemical mutagenesis (specific cases are reviewed in Baranowski et al. 2001, 2003; Jackson et al. 2003; Mason et al. 2003). FMDV mutants produced during replication in cell culture or in animals are under examination and provide valuable tools to help understanding FMDV biology.

Genetic variation of FMDV, like that of other viruses, results in an increasing amount of nucleotide and amino acid sequence information which demands an orderly treatment of data to arrive at a meaningful understanding of the origins of genetic diversity. Genomic sequences have been analyzed by phylogenetic and statistical procedures to define relationships among dominant sequences found in the isolates from infected animals. The result of such analyses is a new genotypic classification of FMDV that is gradually replacing the classic classification based on serology. Phylogenetic studies have also permitted tracing the origin of FMD outbreaks, contributing to the molecular epidemiology of this pathogen (see the chapter by Kitching, this volume). An understanding of the origins of genetic diversity is increasingly achieved by analyzing in detail the genetic population structure of the virus as it replicates in infected hosts, or in cell culture. Probing into the fine structure of viral genomes at the population level has been possible through the application of biological and molecular cloning, combined with rapid nucleotide sequencing.

The population size and genetic heterogeneity of viral populations are increasingly recognized as relevant to viral pathogenesis. Model studies on passage of infectious clones under controlled environments and designed population regimes can provide insight into the major evolutionary forces acting on viruses. For viruses that can produce plaques on cell monolayers (or cause infection from a single viral particle by end-point dilution), the progeny of a single infectious genome can be analysed. In sequence screenings based on biological clones, there is a bias towards scoring genomes which are infectious in the particular cell line chosen, but not in other cells or in some animal hosts. Reverse transcription of genomic RNA and PCR amplification (RT-PCR), followed by molecular cloning and sequencing of individual clones, offers an alternative means to examine viral populations which does not depend on infectivity. Here a bias may come from low fidelity of the enzymes used for RT-PCR (which may result in an overestimate of nucleotide sequence heterogeneity) or from a limitation in the number of viral RNA template molecules amplified by RT-PCR (resulting in an underestimate of nucleotide sequence heterogeneity). Both potential biases can be easily avoided by appropriate control experiments (experimental details for FMDV are given by Arias et al. 2001 and Airaksinen et al. 2003).

Extensive studies with FMDV and other RNA viruses involving the analysis of biological and molecular clones have documented that at the population level viral genomes in infected natural hosts and cell cultures, during acute and persistent infections, consist of complex mutant distributions termed viral quasispecies (Eigen 1971, 1996; Domingo et al. 1978, 2001; Eigen and Schuster 1979; Eigen and Biebricher 1988). This means that the nucleotide sequence determined for a virus isolate (the average or consensus sequence of that particular isolate) often does not exist physically in the population, or exists only as a minority subpopulation within a larger mutant spectrum. The composition of an RNA virus is "statistically defined but individually indeterminate" (Domingo et al. 1978). This population structure is a consequence of the high input of mutations during replication, and it underlies virus adaptability because viruses replicate essentially as pools of genetic and phenotypic variants. The individual components of the mutant spectrum are ranked according to relative fitness (a measure of relative replication capacity in a given environment), and most of them, when they replicate, display lower fitness than the average for the population from which they were isolated (Domingo et al. 1978; Duarte et al. 1994). Fitness, moreover, is unavoidably a property of ensembles of individuals because even virus from a single plaque is a mutant distribution (Escarmís et al. 1996, 2002). Subpopulations of viral genomes provide a reservoir of genomes ready to become the dominant subset in the face of an environmental challenge (e.g., an immune response, the presence of an inhibitor that targets the dominant genome class or an encounter with a new host cell type, among others). The generation of quasispecies swarms is the first stage in the process of genetic diversification of viruses which occurs within infected hosts. Diversification is more clearly manifested upon host-to-host transmission when one or a few founder viruses from a mutant spectrum replicate in a different environment. Further diversification in nature is a complex process, poorly understood in molecular terms, and thought to be influenced by positive selection and random drift of genomes (reviewed in different chapters of Morse 1993, 1994; Gibbs et al. 1995).

Several experiments have shown that virus evolution is directly relevant to the generation of disease or to disease progression within infected individuals. Classic examples are mutations which render attenuated poliovirus virulent and progression to AIDS associated with HIV-1 evolution. In this respect, the elegant work of Kimata et al. (1999) showed that simian immunodeficiency virus molecular clones synthesized from virus isolated from monkeys at different stages of disease reproduced the disease stage of the parental monkey when inoculated into healthy monkeys. Furthermore, many cases of genetic change associated with a modification of host cell tropism and host range with implications in viral pathogenesis have been described (reviewed in Baranowski et al. 2003). Therefore, virus evolution is one of the determinants of viral disease. A common misunderstanding is the thought that many mutations are needed for a substantial biological modification such as an alteration of virus host range or virulence. This is not necessarily the case. One or a few mutations (with a good probability of being represented in mutant spectra of viral quasispecies) may suffice to alter virulence and other important biological traits of viruses. A case involving FMDV is the demonstration that one amino acid replacement selected during replication of a swine virus in guinea-pigs determined the capacity to produce disease in the new host (Núñez et al. 2001). The problem of relating a genetic change to a phenotypic alteration does not arise commonly from the number of mutations but from the fact that several unrelated mutations or multiple unrelated combinations of few mutations may lead to similar phenotypic alterations.

Additional implications of quasispecies dynamics for RNA virus biology have been reviewed recently (Domingo et al. 2001; Domingo 2003) and they are not further discussed here, except with respect to molecular mechanisms of virus extinction and recent trends in the development of new antiviral strategies, the central topics of this chapter.

2 Accumulation of Deleterious Mutations: Rate, Mode, Mechanisms

Several concepts of population genetics have been very useful for the understanding of quasispecies dynamics. One of these is the accumulation of deleterious mutations in asexual populations of organisms when no compensatory mechanisms such as recombination intervene, a process termed Muller's ratchet (Muller 1964; Maynard-Smith 1976). The operation of Muller's ratchet was first documented with an RNA virus by Lin Chao (Chao 1990) working with phage $\phi 6$. These results were then extended to VSV (Duarte et al. 1992), FMDV (Escarmís et al. 1996) and HIV-1 (Yuste et al. 1999). Experimentally, an increase in deleterious mutations and fitness loss can be demonstrated upon serial plaque-to-plaque transfers of virus, in which virus replication is limited to the development of a plaque on the cell monolayer (Fig. 1). Fitness loss associated with accumulation of deleterious mutations is in contrast with fitness gain upon large population passages of virus (Novella et al. 1995) (Fig. 2). In the latter situation, a competitive optimization of the mutant distributions occurs, resulting in selection of mutant distributions that show high fitness in the environment in which replication takes place. The initial fitness of the virus determines whether a given population size involved in replication will lead either to an increase or to a decrease in relative fitness (Novella et al. 1995, 1999).

Nucleotide sequence comparisons carried out with FMDV have defined multiple molecular pathways for fitness loss (Escarmís et al. 1996) or for fitness gain (Escarmís et al. 1999). Remarkable differences were noted in the types of mutations accumulating in the FMDV genome in the course of plaque-to-plaque transfers and those observed among natural isolates, or laboratory populations evolved without bottlenecking. Specifically, in clones derived by serial plaque transfers, 50% of the amino acid substitutions affecting the viral capsid were located at internal

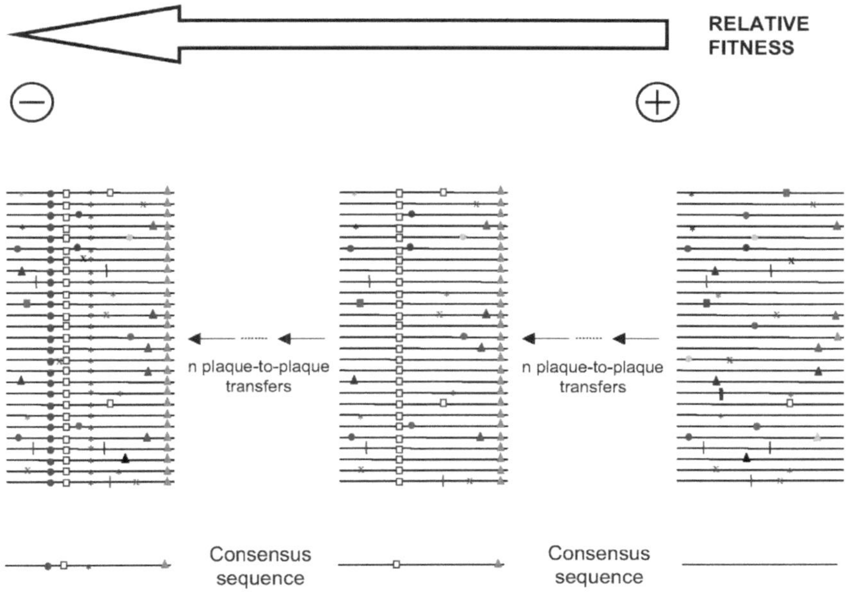

Fig. 1 A scheme of a viral quasispecies and the accumulation of mutations upon plaque-to-plaque transfers. A typical RNA virus mutant distribution is shown on the *right*; genomes are depicted as *horizontal lines* and mutations are represented by different *symbols on the lines*. Plaque-to-plaque transfers (*small arrows*) force the population through successive populations with a modified consensus sequence (*bottom lines*). Note that the consensus sequences are not represented in their respective mutant spectra. Because deleterious mutations tend to be more frequent than neutral and advantageous mutations, relative fitness (*upper arrow*) tends to decrease initially. When fitness is low the dynamics of fitness variation becomes very complex (see text)

capsid sites, which are highly restricted for variation during natural evolution or large population passages of FMDV (Acharya et al. 1989; Mateu et al. 1994). In the course of at least 200 plaque-to-plaque transfers of several viral clones, point mutations accumulated in a nearly linear fashion at a rate of about 0.3 mutations per genome per transfer. This is an average rate of accumulation observed in the consensus sequence of the virus population present in a single plaque, but it does not mean that three plaque transfers are needed for one mutation to occur. During each plaque development multiple mutations occur (as expected from high mutation rates; Sect. 1) and virus from individual plaques is genetically heterogeneous (Escarmís et al. 1996, 2002). Mutations which accu-

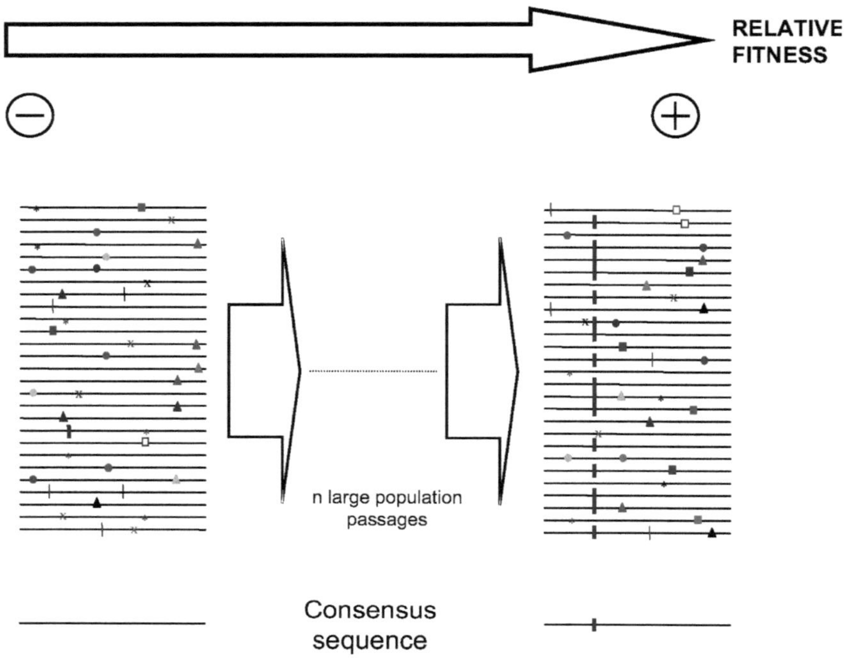

Fig. 2 A scheme of a viral quasispecies and adaptive evolution upon large population passages of the viral population. Symbols to depict viral quasispecies are as in Fig. 1. Large population passages (*thick central arrows*) result in optimization of the mutant distribution and adaptation to the environment in which replication takes place, seen as a change in the consensus (or average sequence) of the population. In this case, relative fitness increases (*upper arrow*). When fitness reaches high values, the population size may become a limiting factor for further increases of fitness, and the population dynamics becomes very complex (see text)

mulated upon plaque transfers of FMDV were clustered at some genomic loci (different for independent lineages) rather than being randomly distributed along the genome. In clusters, the mutation frequencies were four- to five-fold larger than the average for the entire genome, and the difference with the average mutation frequency for the entire genome was statistically significant. The basis for such mutation clustering is not well understood (Escarmís et al. 2002). Possible interpretations include a reduced copying fidelity of the viral replicase at genomic regions affected by template structure or preexisting mutations and the occurrence of compensatory mutations in the proximity of those that are deleterious. In addition, many clonal lineages acquired an elongation of four adenylate residues that precede the second AUG translation initiation codon, which created an internal oligoadenylate, a genetic lesion which had never been observed in FMDV. The oligoadenylate was variable in length, heterogeneous within some viral plaques and associated with fitness loss of FMDV (Escarmís et al. 1996, 1999, 2002). It is one of the genetic markers used to demonstrate the presence of a molecular memory in viral quasispecies (Ruíz-Jarabo et al. 2000, 2002; Arias et al. 2001). The oligoadenylate permitted the experimental demonstration of multiple, alternative pathways for fitness recovery when the debilitated clones were subjected to large population passages. Alternative pathways observed were the shortening of the elongated oligoadenylate which on occasions yielded the wild-type sequence (true reversion) and a deletion of 69 residues spanning the site of the polyadenylate extension (Escarmís et al. 1999). Interestingly, this deletion resulted in viable viruses whose genomes included only 12 nucleotides between the two AUG translation initiation codons (Fig. 3). The mechanism proposed for the generation and elongation or shortening of the internal oligoadenylate tract is polymerase slippage which leads to misalignment mutagenesis (Ripley 1990; Arias et al. 2001). These results with FMDV emphasize how the dynamics of viral infections (e.g., occurrence and severity of genetic bottlenecks, intervention of large population infections) can

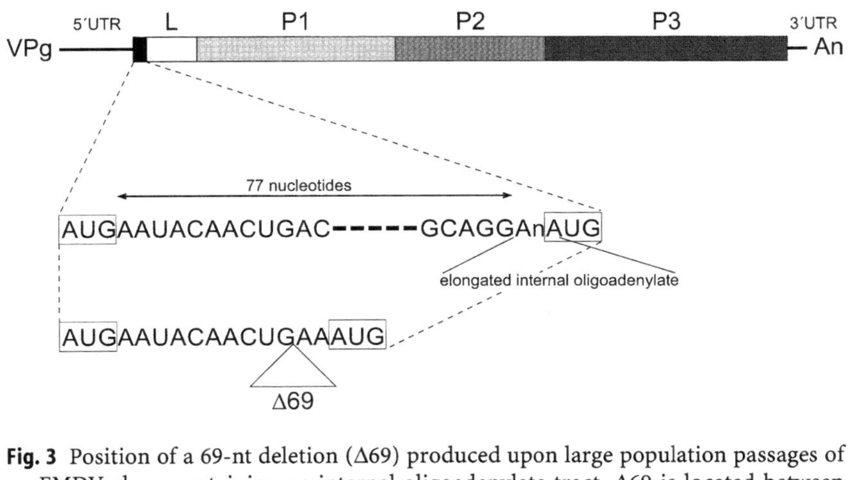

Fig. 3 Position of a 69-nt deletion (Δ69) produced upon large population passages of an FMDV clone containing an internal oligoadenylate tract. Δ69 is located between the two functional AUG protein synthesis initiation codons. The internal oligoadenylate (An) precedes the second functional AUG. Based on results reported in Escarmís et al. 1996, 1999

profoundly affect the types and numbers of mutations observed in a virus genome [the portion of 'sequence space' the mutant spectra occupy (Eigen and Biebricher 1988)] despite using the same replication machinery with its copying fidelity properties, within the same host cells.

3 Resistance to Extinction Despite Accumulation of Mutations: Observations and Modeling

Despite a nearly linear accumulation of mutations in FMDV clones subjected to plaque-to-plaque transfers, the virus showed a remarkable resistance to extinction. An FMDV population (obtained either from a plaque or from an infection in liquid culture medium) is considered extinct when, upon at least three blind passages in cell culture under optimal infection conditions, no infectivity and no FMDV-specific RT-PCR-amplifiable material can be recovered. The amount of infectious virus found in individual plaques developed during a given time period in the same environmental conditions was taken as an approximate measure of relative fitness (Escarmís et al. 2002). Fitness decrease upon serial plaque-to-plaque transfers of several FMDV clones was biphasic. An initial phase of exponential decrease was followed by a second phase in which fitness values displayed large fluctuations around an average constant value; the amplitude of the fluctuations tended to be larger the lower the fitness values (Escarmís et al. 2002; Lázaro et al. 2002) (Fig. 4).

A numerical model was developed which provided clues to a molecular interpretation of the experimental results. A critical feature of the model is the occurrence with low probability of advantageous mutations which permit reaching an equilibrium between the trend to eliminate individuals which have attained very low fitness values and the probability of selecting for the subsequent transfer individuals with compensatory mutations. As a result of both processes a stationary state of constant average fitness is achieved (Lázaro et al. 2002). Nucleotide sequencing of genomic RNA from clones of successive plaque transfers indicated that mutations were associated with fitness fluctuations. In particular, in clones with the internal oligoadenylate, a shortening of the homopolymeric tract, or its interruption by A→G transitions, resulted in fitness increase (Escarmís et al. 2002). The relative fitness was inversely related to the length of the oligoadenylate tract (Fig. 5). Because of these mutation-associated fluctuations in viral production, the FMDV clones displayed a remarkable resistance to extinction. Here a constant muta-

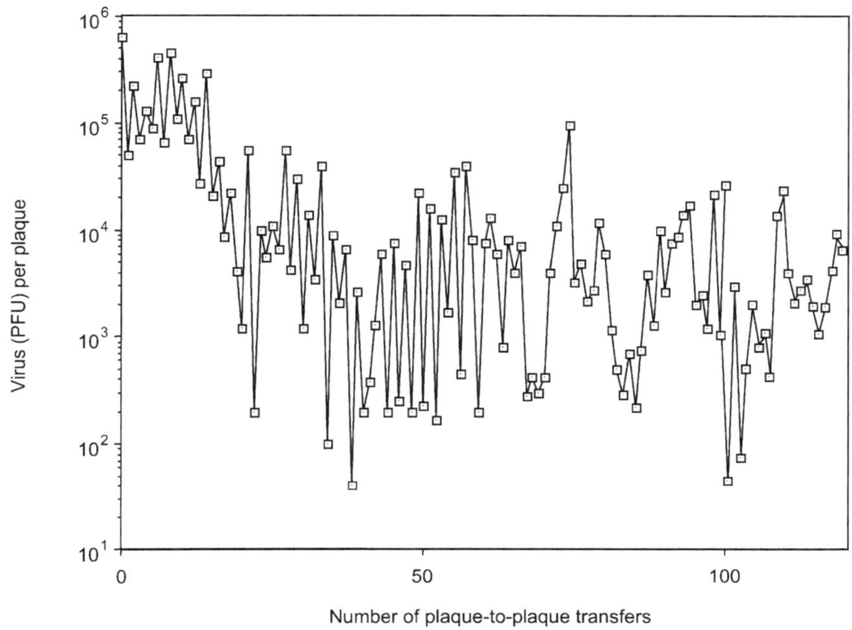

Fig. 4 Evolution of relative fitness of an FMDV clone subjected to plaque-to-plaque transfers. At each plating on a BHK-21 cell monolayer, a plaque (developed for 24 h under a semisolid agar overlay) was chosen at random and the infectious virus present in it was determined by a plaque assay (values on *ordinate*). Virus from the same plaque was plated again, and the process was repeated a total of 120 times (*abcissa*). A biphasic evolution and large fluctuations of fitness values can be observed. Similar findings were obtained with additional FMDV clones. Based on results reported and discussed in Escarmís et al. 2002; Lázaro et al. 2002, 2003

tional input serves to attenuate the effects of Muller's ratchet to render them unnoticeable regarding virus survival. Manrubia and associates (Manrubia et al. 2003) have explored a mean field theory which in fact represents an extension of the model developed for FMDV to any exponentially growing population of mutating replicons. The mean field theory proposed permits an exact formal analysis of various dynamical regimes. When the population is subjected to strong bottlenecks, large fluctuations of viral yield at each passage are expected, but they do not drive the population to extinction. A precise mathematical form describes the fluctuations as observed experimentally. When the population is allowed to grow exponentially for many replication cycles, the relative proportion of different mutant types attains fixed values. Remark-

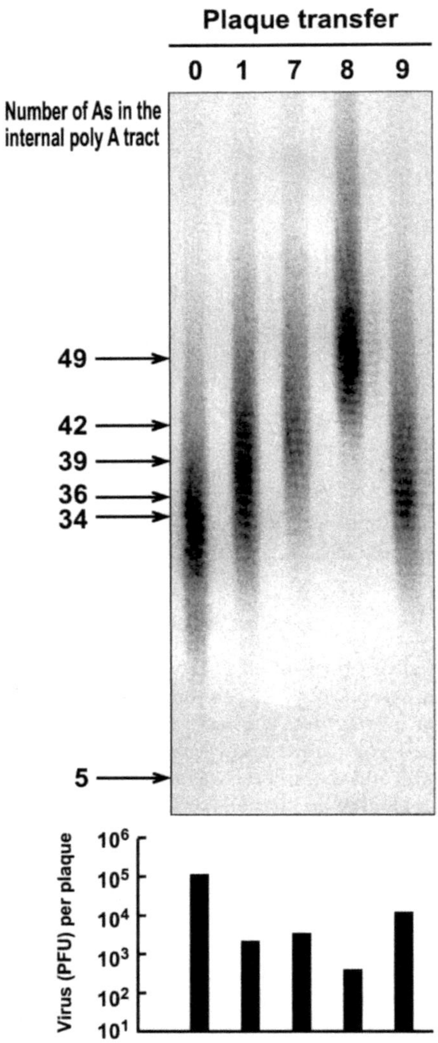

Fig. 5 Length of the internal oligoadenylate tract and infectivity of FMDV clones. The genomic region spanning the internal oligoadenylate tract (compare with Fig. 3) was amplified by RT-PCR in the presence of S^{35}-dATP. The amplified products (a 198-bp amplicon) were analyzed by electrophoresis through a 6% polyacrylamide sequence gel (*top of figure*). The number of PFU contained in the corresponding plaques was also determined (*bottom of figure*). This and other results documented that a longer oligoadenylate implies a lower infectivity. Based on experiments reported in Escarmís et al. 2002

ably and unexpectedly, in this model, for high enough mutation rates, the most represented variant is not the one with the highest fitness (Manrubia et al. 2003).

In a further theoretical development linked to the results on repeated bottlenecking of FMDV, Lázaro et al. (2003) have shown that the fluctuating pattern of fitness values of FMDV clones follows a Weibull distribution. This is a type of statistical distribution (Weibull 1951) which describes unrelated processes such as fragmentation of materials, cardiac contractions and time between events of the disease paroxysmal atrial fibrillations (Araki et al. 1999; Rose et al. 1999). This pattern essentially reflects the large variations in the initial fitness of the founder virus at each bottleneck event, which are exponentially enhanced during plaque development. Variations in fitness have their origin in the mutations occurring in the viral genome during plaque development and result from the extreme complexity of the host-virus interaction, involving multiple viral and host cell functions (Domingo 2003; Lázaro et al. 2003). As the virus becomes more debilitated by deleterious mutations, compensatory mutations play a more relevant role to increase the fitness of the virus, contributing in this way to the fluctuating pattern of fitness values and survival of genome subsets from the population (Lázaro et al. 2002, 2003).

Concepts that find application to virology may arise as an integration of results from experimental virology with population genetics and physics. One such concept is complexity, which refers to systems made of multiple elements, whose properties (or behavior) cannot be anticipated merely as the sum of contributions of the individual elements which comprise the system (Gell-Mann 1994). FMDV, like other RNA viruses, has a compact genetic information in its genome (Sect. 1), so that many genomic regions are involved in multiple functions. Therefore, mutations, so frequent in each replication round, when accumulated as the result of bottlenecks, may trigger a cascade of perturbations which lead to large variations in the fitness of the individual components of a lytic plaque. That this is so (versus the alternative of accumulation of mutations which sum small biological effects) has been unveiled by the numerical analyses of plaque-to-plaque virus yield values which have shown that the system follows a Weibull distribution (Lázaro et al. 2003). The effect of mutations, however complex and continuous, did not drive the entire FMDV population to extinction. Alternative pathways of accumulation of mutations which lead to effective extinction of viruses must be sought.

4
Lethal Mutagenesis or the Transition into Error Catastrophe: A Pathway Towards Virus Extinction

A consequence of error-prone replication and quasispecies dynamics is the existence of an error threshold for maintenance of genetic information. This important concept was proposed on a theoretical basis (Swetina and Schuster 1982; Eigen and Biebricher 1988; Nowak and Schuster 1989; Alves and Fontanari 1998; Nilsson and Snoad 2000; Eigen 2002), and has been supported by an increasing number of experimental results with different viruses (Holland et al. 1990; Loeb et al. 1999; Crotty et al. 2000, 2001; Loeb and Mullins 2000; Lanford et al. 2001; Grande-Pérez et al. 2002; Domingo et al. 2003; Severson et al. 2003). The first experimental evidence of a transition of virus into error catastrophe was obtained by J.J. Holland and his colleagues, who documented the adverse effect of chemical mutagenesis on VSV and poliovirus (PV) (Holland et al. 1990; Lee et al. 1997). These results suggested that VSV and PV replicate close to the error threshold. Increases in mutation rate should result in violation of the error threshold and an irreversible transition from a productive viral infection into an abortive viral infection, reflected in virus extinction. It must be emphasized that virus entry into error catastrophe implies virus extinction, not merely a decrease in virus titer.

Experiments with FMDV have been designed to evaluate the effect of relative viral fitness and viral load (the number of infectious particles that participate in the infection) in virus extinction by mutagenic base analogs. For a given viral load, a low relative fitness favors virus extinction (Sierra et al. 2000; Pariente et al. 2001). This satisfies a prediction of quasispecies theory because a lower viral fitness implies a lower superiority of the master sequence in the quasispecies distribution, and this is one of the parameters which determine complexity of the genetic information that can be maintained (the position of the error threshold) (Swetina and Schuster 1982; Nowak and Schuster 1989; Alves and Fontanari 1998; Nilsson and Snoad 2000; Eigen 2002). For a given relative fitness, the lower the viral load, the more likely it was to drive FMDV to extinction by a given dose of mutagenic agent (Sierra et al. 2000; Pariente et al. 2001). These observations led to investigations of FMDV extinction by the combined action of the mutagenic base analog 5-fluorouracil (FU) and the antiviral inhibitors guanidine hydrochloride and heparin. Some of these treatments led to systematic extinction (47 times out of 47 attempts) of high-fitness FMDV in fewer than five passages in cell culture (Pariente et al. 2001). Interestingly, extinction of high-fitness FMDV was achieved in combinations which included FU but not with guanidine hydrochloride and heparin alone, even when these two inhibitors exerted the same inhibitory effect as FU in one passage (Airaksinen et al. 2003). These results suggest that combinations of viral-specific mutagenic agents and antiviral inhibitors could provide suitable treatments to eradicate virus from infected organisms.

The FMDV model also provided an experimental system to approach the problem of virus escape due to resistance to antiviral inhibitors, an evolutionary phenomenon which greatly impairs treatment by antiviral inhibitors of important diseases such as AIDS, hepatitis C or influenza (Richman 1996; Domingo et al. 2001; Menéndez-Arias 2002). A detailed analysis of the FMDV populations which escaped extinction showed that in all cases mutations that confer resistance to guanidine hydrochloride and/or heparin were selected (Airaksinen et al. 2003). In contrast, the consensus sequence of preextinction populations (those preceding extinction) did not show any mutation relative to the wild type. In the process towards extinction by a combination of FU and guanidine hydrochloride, there was a 1,000-fold reduction in virus titer and a 100-fold reduction in the amount of viral RNA (Pariente 2003). This and other observations (C. González-López and N. Pariente, unpublished results) suggest that there is a considerable amount of non-infectious, mutagenized RNA in the process of transition to error catastrophe (Airaksinen et al. 2003). These results reinforce the conclusions that mutagenesis rather than inhibition of viral-specific RNA synthesis is the main mechanism of action of FU in the extinction of FMDV. Similar conclusions have been reached in studies with the prototype arenavirus lymphocytic choriomeningitis virus (LCMV) (Grande-Pérez et al. 2002; Ruiz-Jarabo et al. 2003). Intracellular UTP pools were depleted in FU-treated cells, whereas there was an accumulation of fluorouridine triphosphate (FUTP), resulting in approximately 12 times more intracellular FUTP than UTP. The other NTPs remained close to normal levels (Airaksinen et al. 2003). These measurements strongly suggest that FU-mediated mutagenesis of FMDV is associated with incorporation of fluorouridine monophosphate into viral RNA rather than with nucleotide pool imbalances. Experiments are in progress to investigate this point further.

The transition into error catastrophe was accompanied by increases in mutant spectrum complexity as measured by mutation frequencies (proportion of mutated positions) and Shannon entropies (proportion of non-identical genomes). Such increases varied depending on the virus (poliovirus, HIV-1, FMDV or LCMV), the genomic region analyzed, and the mutagenic treatment (Loeb et al. 1999; Sierra et al. 2000; Crotty et al. 2001; Pariente et al. 2001; Grande-Pérez et al. 2002). In the case of FMDV, mutation frequencies in the mutant spectra of populations subjected to FU mutagenesis were 1.5- to 6-fold larger than for control populations subjected to parallel passages in the absence of mutagen. The maximum relative increases in complexity were seen in the polymerase (3D)-coding region, a genomic region which, under unperturbed conditions of replication, is highly conserved and which usually displays a very narrow mutant spectrum (Sierra et al. 2000). Despite increases in mutant spectrum complexity, consensus sequences in preextinction FMDV populations did not vary with respect to the parental virus, suggesting that mutagenic pressure impairs adaptive selection of mutant distributions with new dominant sequences. In contrast, genomic consensus sequences of FMDV populations which were subjected to serial passages in the presence of mutagenic agents, but did not result in virus extinction, showed mutations in the consensus sequence with respect to the parental virus (Sierra et al. 2000). This result suggests that in this case adaptive mutations occurred which contributed to elude extinction, as also observed with extinction-escape mutants of FMDV harboring inhibitor-resistant mutations when mutagenesis was insufficient to achieve virus extinction (Airaksinen et al. 2003).

Ribavirin (1-β-D-ribofuranosyl-1,2,3-triazole-3-carboxamide) is a broad-spectrum nucleoside analog which can exert its antiviral activity through several mechanisms (Snell 2001; Zhang et al. 2003). Interestingly, this licensed drug has been shown to be mutagenic for a number of RNA viruses (Crotty et al. 2000, 2001; Lanford et al. 2001; Maag et al. 2001; Airaksinen et al. 2003; Severson et al. 2003). Ribavirin triphosphate (RTP) can be incorporated by RNA-dependent RNA polymerases (Crotty et al. 2000, 2001; Maag et al. 2001), thus providing a molecular interpretation of its mutagenic action. Recently, a poliovirus polymerase mutant with decreased sensitivity to ribavirin has been characterized (Pfeiffer and Kirkegaard 2003), emphasizing the need to cross the error threshold to avoid selection not only of virus mutants resistant to inhibitors which may be used in combination with mutagens (Airaksinen et al. 2003) but also of mutants which may show decreased sensitivity to mutagenic agents. However, mutagenesis is not the only mechanism by which ribavirin exerts its antiviral activity. Its potent anti-arenavirus activity was not associated with significant increases in mutant spectrum complexity of LCMV (Ruiz-Jarabo et al. 2003), a result which is in contrast with the effects of FU in the same system (Grande-Pérez et al. 2002; Ruiz-JARABO et al. 2003). Application of microarray-based detection of perturbations in cellular gene expression documented multiple alterations of gene expression which were associated with administration of ribavirin in cells infected with respiratory syncytial virus (Zhang et al. 2003).

Ribavirin can eliminate FMDV from persistently infected cell cultures (de la Torre et al. 1987). To examine whether this curing activity was associated with enhanced mutagenesis, the alterations of mutant spectra complexity of persistent FMDV as a result of treatment with ribavirin or with mycophenolic acid were compared (Airaksinen et al. 2003). Both drugs are inhibitors of inosine monophosphate dehydrogenase (IMPDH), a key enzyme of the pathway of the synthesis of GTP, and therefore administration of either of these two drugs results in similar intracellular nucleotide pool imbalances. The critical difference between the two inhibitors is that RTP can be a substrate for viral polymerases whereas mycophenolic acid cannot be incorporated into polynucleotide chains. Mycophenolic acid-induced mutagenesis of FMDV was weak and reversed by addition of guanosine to the culture medium, and thus it was probably due to nucleotide pool alterations. In contrast, ribavirin addition resulted in higher mutation frequencies (Airaksinen et al. 2003) which reached levels comparable to those observed in other viral systems (Crotty et al. 2000, 2001; Contreras et al. 2002; Severson et al. 2003; Zhou et al. 2003). The antiviral effect of ribavirin was diminished by guanosine addition, but its mutagenic activity was not. These observations suggest that ribavirin-mediated curing of FMDV from persistently infected cultures was associated with at least two effects: mutagenesis and inhibition of IMPDH (Airaksinen et al. 2003). Despite ribavirin having multiple mechanisms of action, the demonstration that at least in some cases its antiviral activity may be associated with a mutagenic activity has represented important progress in the prospects of developing error catastrophe as a new antiviral strategy (Graci and Cameron 2002).

de la Torre and colleagues (Ruiz-Jarabo et al. 2003) have shown that treatment of mice with FU prevents the establishment of a persistent infection with LCMV. This important result constitutes a proof of principle of the feasibility of a mutagenesis-based, antiviral approach in vivo.

5
Advantages and Limitations of Error Catastrophe as an Antiviral Strategy

A comparison of the results with FMDV showing the robust resistance of this virus to extinction when mutations accumulate during bottleneck transfers (Escarmís et al. 2002; Lázaro et al. 2002, 2003), with results showing extinction by mutagenic treatments (Sierra et al. 2000; Pariente et al. 2001; Airaksinen et al. 2003; Pariente 2003) points to modulation of the mutagenic input as critical for virus survival. This does not mean that upon serial bottleneck transfers extinction events are rare. They are probably frequent, but the mutant repertoire allows for a viable virus (in this case 'viable' means 'able to form a plaque') to arise. An excessive mutation rate suppresses this 'viability-escape' potential of the mutant reservoir. The critical contribution of mutation rates to extinction, predicted by theoretical treatments (Swetina and Schuster 1982; Eigen and Biebricher 1988; Nowak and Schuster 1989; Alves and Fontanari 1998; Nilsson and Snoad 2000) has been documented experimentally with the comparative studies with FMDV (Sects. 3 and 4). The irreversibility of the transition provides a definitive advantage of error catastrophe versus inhibition as an antiviral design (Airaksinen et al. 2003; C. González-López et al., manuscript in preparation}.

Despite the promising developments summarized in Section 4, a clinical application of lethal mutagenesis requires addressing a number of issues, including (a) the specificity for viral replicases and retrotranscriptases of the antiviral mutagens to be used alone or in combination with antiviral inhibitors. This is a key point which requires the development and testing of new mutagenic agents which can be incorporated by viral enzymes but not by cellular enzymes. (It must be said, however, that many drugs which have been licensed as antiviral agents show considerable toxicity for cells and organisms). (b) A second problem to be addressed is that the concentration of mutagenic agent(s) at the sites of viral replication must be sufficiently high to provoke virus replication to cross the error threshold. Insufficient mutagenesis could favor survival of virus mutants with unpredictable biological properties, or even mutants manifesting resistance to mutagenic agents (Pfeiffer and Kirkegaard 2003). (c) To limit possible resistance to lethal mutagenesis it may be necessary to use more than one mutagen simultaneously to avoid viral escape (as with combination therapy with inhibitors). (d) In the case of retroviruses, the integrated provirus will be largely immune to mutagenic treatments because proviral DNA is copied by the cellular replication machinery as though it were a cellular gene. Retrovirus entry into error catastrophe may necessitate combining activation of provirus to enter the particle formation pathway together with mutagenic treatments.

These limitations for an eventual clinical application of error catastrophe are not substantially different to those limitations often encountered with current antimicrobial treatments, and they should encourage rather than discourage research on antiviral treatments aimed at virus extinction through enhanced mutagenesis. This new line of research is favored by an increasing understanding of the molecular basis of polymerase copying fidelity (Menéndez-Arias 2002), as well as by expanding possibilities for the design of new mutagenic base analogs, useful also for other types of medical interventions such as cancer therapy. When low toxicity and means to reach effective mutagen or mutagen-inhibitor combinations are achieved, the chances of success will be high. This great potential will only be realized as a result of gradual accumulation of basic information by studying multiple virus-host systems.

6
Conclusions and Prospects

In this article we have summarized experimental and theoretical approaches to the understanding of quasispecies dynamics of FMDV aimed at defining parameters which can shift virus populations from sustained survival to extinction. One of the advantages of exploiting evolutionary concepts to design antiviral strategies is that developments can benefit from studies with multiple virus-host systems because the molecular bases of the designs are shared by all viral systems. One of the common threads which underlie quasispecies dynamics—on which error catastrophe methodology is constructed—is error-prone replication. Low fidelity of template copying is expected to occur whenever RNA is the genetic material (in all riboviruses and retroviruses) or RNA is a replicative intermediate in DNA viruses (in hepadnaviruses such as hepatitis B virus). Even if some proofreading activity operated during replication of the largest RNA genomes—for example, an activity of the type described for human influenza virus polymerase (Ishihama et al. 1986)—it is unlikely that post-replicative correction pathways would be efficient on those RNA genomes. Therefore, high mutation rates, population heterogeneity, quasispecies dynamics and the potential for rapid evolution in nature stand as general features of RNA viruses and probably also of some DNA viruses (Domingo 2003). These general features encourage equally general antiviral designs such as lethal mutagenesis.

There are precedents in biological systems of exploitation of enhanced mutagenesis as a defense mechanism of cells against invading molecular parasites. Filamentous fungi such as *Neurospora crassa* have evolved a mechanism against DNA with repeated nucleotide sequences which can penetrate into their cells, consisting in producing mutations at each of the repeat copies. This mechanism is known as repeated-induced point mutations (RIP) (Kinsey et al. 1994; reviews in Bushman 2002 and Arnold and Hilton 2003). The existence of an innate cellular immunity to retroviral infections has recently been described (Harris et al. 2003; Lecossier et al. 2003; Mangeat et al. 2003; Zhang et al. 2003). This cellular defense is mediated by mutagenesis of the viral genome through cytidine deamination which greatly impairs expansion of retroviruses. Protein vif of HIV-1 overcomes this mutagenesis effect through a still -unknown mechanism. Therefore, a mutagenesis-based antiviral approach to drive virus to extinction may not be foreign to the natural mechanisms which have permitted survival of organisms in the face of perturbing molecular parasites.

Acknowledgements Work at CBMSO was supported by grants BMC 2001.1823C02-01 from MCyT, 08.2/0015/2001 from CAM, and Fundación Ramón Areces. Work at CAB was supported by the EU, CAMP and INTA. NP acknowledges a predoctoral fellowship from MCyT and a postgraduate fellowship from CSIC; AA a Marie Curie Fellowship of the European Community Quality of Life and Management of Living Resources program (contract QLK-CT-1999-51462); SCM a Contrato Ramón y Cajal from MCyT; CG-L a postdoctoral fellowship from CAM; and MH a predoctoral fellowship from MCyT.

References

Acharya R, Fry E, Stuart D, Fox G, Rowlands D, Brown F (1989) The three-dimensional structure of foot-and-mouth disease virus at 2.9 Å resolution. Nature 337:709–716

Airaksinen A, Pariente N, Menendez-Arias L, Domingo E (2003) Curing of foot-and-mouth disease virus from persistently infected cells by ribavirin involves enhanced mutagenesis. Virology 311:339–349

Alves D, Fontanari JF (1998) Error threshold in finite populations. Phys Rev E 57:7008–7013

Araki J, Matsubara H, Shimizu J, Mikane T, Mohri S, Mizuno J, Takaki M, Ohe T, Hirakawa M, Suga H (1999) Weibull distribution function for cardiac contraction: integrative analysis. Am J Physiol 277:H1940–H1945

Arias A, Lázaro E, Escarmís C, Domingo E (2001) Molecular intermediates of fitness gain of an RNA virus: characterization of a mutant spectrum by biological and molecular cloning. J Gen Virol 82:1049–1060

Arnold J, Hilton N (2003) Genome sequencing: revelations from a bread mould. Nature 422:821–822

Bachrach HL (1968) Foot-and-mouth disease virus. Annu Rev Microbiol 22:201–244

Baranowski E, Ruíz-Jarabo CM, Lim P, Domingo E (2001) Foot-and-mouth disease virus lacking the VP1 G-H loop: the mutant spectrum uncovers interactions among antigenic sites for fitness gain. Virology 288:192–202

Baranowski E, Ruíz-Jarabo CM, Pariente N, Verdaguer N, Domingo E (2003) Evolution of cell recognition by viruses: a source of biological novelty with medical implications. Adv Virus Res 62:19–111

Bushman F (2002) Lateral DNA Transfer. Mechanisms and Consequences. Cold Spring Harbor, New York, Cold Spring Harbor Laboratory Press

Chao L (1990) Fitness of RNA virus decreased by Muller's ratchet. Nature 348:454–455

Contreras AM, Hiasa Y, He W, Terella A, Schmidt EV, Chung RT (2002) Viral RNA mutations are region specific and increased by ribavirin in a full-length hepatitis C virus replication system. J Virol 76:8505–8517

Crotty S, Cameron CE, Andino R (2001) RNA virus error catastrophe: direct molecular test by using ribavirin. Proc Natl Acad Sci USA 98:6895–6900.

Crotty S, Maag D, Arnold JJ, Zhong W, Lau JYN, Hong Z, Andino R, Cameron CE (2000) The broad-spectrum antiviral ribonucleotide, ribavirin, is an RNA virus mutagen. Nat Med 6:1375–1379

de la Torre JC, Alarcón B, Martínez-Salas E, Carrasco L, Domingo E (1987) Ribavirin cures cells of a persistent infection with foot-and-mouth disease virus in vitro. J Virol 61:233–235

Doel TR (2003) FMD vaccines. Virus Res 91:81–99

Domingo E, Ed. (2003) Microbe-host interaction: viruses. Curr Op Microbiol 6:4

Domingo E, Baranowski E, Escarmís C, Sobrino F, Holland JJ (2002). Error frequencies of picornavirus RNA polymerases: evolutionary implications for viral populations Molecular Biology of Picornaviruses. Semler BL Wimmer E. Washington, DC, ASM: 285–298

Domingo E, Biebricher C, Eigen M, Holland JJ (2001) Quasispecies and RNA Virus Evolution: Principles and Consequences. Austin, Landes Bioscience

Domingo E, Escarmis C, Baranowski E, Ruiz-Jarabo CM, Carrillo E, Nunez JI, Sobrino F (2003) Evolution of foot-and-mouth disease virus. Virus Res 91:47–63

Domingo E, Sabo D, Taniguchi T, Weissmann C (1978) Nucleotide sequence heterogeneity of an RNA phage population. Cell 13:735–744

Domingo E, Verdaguer N, Ochoa WF, Ruiz-Jarabo CM, Sevilla N, Baranowski E, Mateu MG, Fita I (1999) Biochemical and structural studies with neutralizing antibodies raised against foot-and-mouth disease virus. Virus Res 62:169–175

Drake JW, Holland JJ (1999) Mutation rates among RNA viruses. Proc Natl Acad Sci USA 96:13910–13913

Duarte E, Clarke D, Moya A, Domingo E, Holland J (1992) Rapid fitness losses in mammalian RNA virus clones due to Muller's ratchet. Proc Natl Acad Sci USA 89:6015–6019

Duarte EA, Novella IS, Ledesma S, Clarke DK, Moya A, Elena SF, Domingo E, Holland JJ (1994) Subclonal components of consensus fitness in an RNA virus clone. J Virol 68:4295–4301
Eigen M (1971) Self-organization of matter and the evolution of biological macromolecules. Naturwissenschaften 58:465–523
Eigen M (1996) On the nature of virus quasispecies. Trends Microbiol 4:216–218
Eigen M (2002) Error catastrophe and antiviral strategy. Proc Natl Acad Sci USA 99:13374–13376.
Eigen M, Biebricher CK (1988). Sequence space and quasispecies distribution. RNA Genetics. Domingo E, Ahlquist P Holland JJ. Boca Raton, FL., CRC Press. 3: 211–245
Eigen M, Schuster P (1979) The hypercycle. A principle of natural self-organization. Berlin, Springer
Escarmís C, Dávila M, Charpentier N, Bracho A, Moya A, Domingo E (1996) Genetic lesions associated with Muller's ratchet in an RNA virus. J Mol Biol 264:255–267
Escarmís C, Dávila M, Domingo E (1999) Multiple molecular pathways for fitness recovery of an RNA virus debilitated by operation of Muller's ratchet. J Mol Biol 285:495–505
Escarmís C, Gómez-Mariano G, Dávila M, Lázaro E, Domingo E (2002) Resistance to extinction of low fitness virus subjected to plaque-to-plaque transfers: diversification by mutation clustering. J Mol Biol 315:647–661.
Gell-Mann M (1994). Complex adaptive systems. Complexity Metaphors, models and reality. Cowan GA, Pines D Meltzer D. Reading, MA, Wesley Publishing Co.: 17–45
Gibbs A, Calisher C, García-Arenal F, Eds. (1995) Molecular Basis of Virus Evolution. Cambridge, Cambridge University Press
Graci JD, Cameron CE (2002) Quasispecies, error catastrophe, and the antiviral activity of ribavirin. Virology 298:175–180.
Grande-Pérez A, Sierra S, Castro MG, Domingo E, Lowenstein PR (2002) Molecular indetermination in the transition to error catastrophe: systematic elimination of lymphocytic choriomeningitis virus through mutagenesis does not correlate linearly with large increases in mutant spectrum complexity. Proc Natl Acad Sci USA 99:12938–12943.
Harris RS, Bishop KN, Sheehy AM, Craig HM, Petersen-Mahrt SK, Watt IN, Neuberger MS, Malim MH (2003) DNA deamination mediates innate immunity to retroviral infection. Cell 113:803–809
Holland J, Spindler K, Horodyski F, Grabau E, Nichol S, VandePol S (1982) Rapid evolution of RNA genomes. Science 215:1577–1585
Holland JJ, Domingo E, de la Torre JC, Steinhauer DA (1990) Mutation frequencies at defined single codon sites in vesicular stomatitis virus and poliovirus can be increased only slightly by chemical mutagenesis. J Virol 64:3960–3962
Ishihama A, Mizumoto K, Kawakami K, Kato A, Honda A (1986) Proofreading function associated with the RNA-dependent RNA polymerase from influenza virus. J Biol Chem 261:10417–10421
Jackson T, King AM, Stuart DI, Fry E (2003) Structure and receptor binding. Virus Res 91:33–46

Kimata JT, Kuller L, Anderson DB, Dailey P, Overbaugh J (1999) Emerging cytopathic and antigenic simian immunodeficiency virus variants influence AIDS progression. Nat Med 5:535–541

Kinsey JA, Garrett-Engele PW, Cambareri EB, Selker EU (1994) The Neurospora transposon Tad is sensitive to repeat-induced point mutation (RIP). Genetics 138:657–664

Lanford RE, Chavez D, Guerra B, Lau JY, Hong Z, Brasky KM, Beames B (2001) Ribavirin induces error-prone replication of GB virus B in primary tamarin hepatocytes. J Virol 75:8074–8081

Lázaro E, Escarmís C, Domingo E, Manrubia SC (2002) Modeling viral genome fitness evolution associated with serial bottleneck events: evidence of stationary states of fitness. J Virol 76:8675–8681.

Lázaro E, Escarmís C, Pérez-Mercader J, Manrubia SC, Domingo E (2003) Resistance of virus to extinction upon bottleneck passages: study of a decaying and fluctuating pattern of fitness loss. Proc Natl Acad Sci USA 100:10830–10835

Lecossier D, Bouchonnet F, Clavel F, Hance AJ (2003) Hypermutation of HIV-1 DNA in the absence of the Vif protein. Science 300:1112

Lee CH, Gilbertson DL, Novella IS, Huerta R, Domingo E, Holland JJ (1997) Negative effects of chemical mutagenesis on the adaptive behavior of vesicular stomatitis virus. J Virol 71:3636–3640

Loeb LA, Essigmann JM, Kazazi F, Zhang J, Rose KD, Mullins JI (1999) Lethal mutagenesis of HIV with mutagenic nucleoside analogs. Proc Natl Acad Sci USA 96:1492–1497

Loeb LA, Mullins JI (2000) Lethal mutagenesis of HIV by mutagenic ribonucleoside analogs. AIDS Res Hum Retroviruses 13:1–3

Maag D, Castro C, Hong Z, Cameron CE (2001) Hepatitis C virus RNA-dependent RNA polymerase (NS5B) as a mediator of the antiviral activity of ribavirin. J Biol Chem 276:46094–46098

Mangeat B, Turelli P, Caron G, Friedli M, Perrin L, Trono D (2003) Broad antiretroviral defence by human APOBEC3G through lethal editing of nascent reverse transcripts. Nature 424:99–103

Manrubia SC, Lázaro E, Pérez-Mercader J, Escarmís C, Domingo E (2003) Fitness distributions in exponentially growing asexual populations. Phys Rev Lett 90: 188101–188104

Mason PW, Grubman MJ, Baxt B (2003) Molecular basis of pathogenesis of FMDV. Virus Res 91:9–32

Mateu MG, Hernández J, Martínez MA, Feigelstock D, Lea S, Pérez JJ, Giralt E, Stuart D, Palma EL, Domingo E (1994) Antigenic heterogeneity of a foot-and-mouth disease virus serotype in the field is mediated by very limited sequence variation at several antigenic sites. J Virol 68:1407–1417

Maynard-Smith J (1976) The Evolution of Sex. Cambridge, Cambridge University Press

Menéndez-Arias L (2002) Molecular basis of fidelity of DNA synthesis and nucleotide specificity of retroviral reverse transcriptases. Prog Nucl Acid Res Mol Biol 71:91–147

Menéndez-Arias L (2002) Targeting HIV: antiretroviral therapy and development of drug resistance. Trends Pharmacol Sc 23:381–388

Morse SS, Ed. (1993) Emerging viruses, Oxford, Oxford University Press
Morse SS, Ed. (1994) The Evolutionary Biology of Viruses. New York, Raven Press
Muller HJ (1964) The relation of recombination to mutational advance. Mut Res 1:2–9
Nilsson M, Snoad N (2000) Error threshold for quasispecies in dynamics fitness landscapes. Phys Rev Lett 84:191–194
Novella IS, Duarte EA, Elena SF, Moya A, Domingo E, Holland JJ (1995) Exponential increases of RNA virus fitness during large population transmissions. Proc Natl Acad Sci USA 92:5841–5844
Novella IS, Elena SF, Moya A, Domingo E, Holland JJ (1995) Size of genetic bottlenecks leading to virus fitness loss is determined by mean initial population fitness. J Virol 69:2869–2872
Novella IS, Quer J, Domingo E, Holland JJ (1999) Exponential fitness gains of RNA virus populations are limited by bottleneck effects. J Virol 73:1668–1671
Nowak M, Schuster P (1989) Error thresholds of replication in finite populations mutation frequencies and the onset of Muller's ratchet. J Theor Biol 137:375–395.
Núñez JI, Baranowski E, Molina N, Ruiz-Jarabo CM, Sánchez C, Domingo E, Sobrino F (2001) A single amino acid substitution in nonstructural protein 3A can mediate adaptation of foot-and-mouth disease virus to the guinea pig. J Virol 75:3977–3983
Pariente N (2003). Base molecular y dinámica de la extinción del virus de la fiebre aftosa por combinación de un mutágeno e inhibidores. Ph. D. Thesis, Universidad Autónoma de Madrid.
Pariente N, Sierra S, Lowenstein PR, Domingo E (2001) Efficient virus extinction by combinations of a mutagen and antiviral inhibitors. J Virol 75:9723–9730.
Pfeiffer JK, Kirkegaard K (2003) A single mutation in poliovirus RNA-dependent RNA polymerase confers resistance to mutagenic nucleotide analogs via increased fidelity. Proc Natl Acad Sci USA
Richman DD, Ed. (1996) Antiviral Drug Resistance. New York, John Wiley and Sons Inc.
Ripley LS (1990) Frameshift mutation: determinants of specificity. Annu Rev Genetics 24:189–213
Rose MS, Gillis AM, Sheldon RS (1999) Evaluation of the bias in using the time to the first event when the inter-event intervals have a Weibull distribution. Stat Med 18:139–154
Ruíz-Jarabo CM, Arias A, Baranowski E, Escarmís C, Domingo E (2000) Memory in viral quasispecies. J Virol 74:3543–3547
Ruíz-Jarabo CM, Arias A, Molina-París C, Briones C, Baranowski E, Escarmís C, Domingo E (2002) Duration and fitness dependence of quasispecies memory. J Mol Biol 315:285–296
Ruiz-Jarabo CM, Ly C, Domingo E, Torre JC (2003) Lethal mutagenesis of the prototypic arenavirus lymphocytic choriomeningitis virus (LCMV). Virology 308:37–47
Severson WE, Schmaljohn CS, Javadian A, Jonsson CB (2003) Ribavirin causes error catastrophe during Hantaan virus replication. J Virol 77:481–488

Sierra S, Dávila M, Lowenstein PR, Domingo E (2000) Response of foot-and-mouth disease virus to increased mutagenesis. Influence of viral load and fitness in loss of infectivity. J Virol 74:8316–8323

Snell NJ (2001) Ribavirin—current status of a broad spectrum antiviral agent. Expert Opin Pharmacother 2:1317–1324

Sutmoller P, Barteling SS, Olascoaga RC, Sumption KJ (2003) Control and eradication of foot-and-mouth disease. Virus Res 91:101–144

Swetina J, Schuster P (1982) Self-replication with errors. A model for polynucleotide replication. Biophys Chem 16:329–345

Weibull WJ (1951) A statistical distribution function of wide applicability. Appl Mech 18:293–297

Yuste E, Sánchez-Palomino S, Casado C, Domingo E, López-Galíndez C (1999) Drastic fitness loss in human immunodeficiency virus type 1 upon serial bottleneck events. J Virol 73:2745–2751

Zhang H, Yang B, Pomerantz RJ, Zhang C, Arunachalam SC, Gao L (2003) The cytidine deaminase CEM15 induces hypermutation in newly synthesized HIV-1 DNA. Nature 424:94–98

Zhang Y, Jamaluddin M, Wang S, Tian B, Garofalo RP, Casola A, Brasier AR (2003) Ribavirin treatment up-regulates antiviral gene expression via the interferon-stimulated response element in respiratory syncytial virus-infected epithelial cells. J Virol 77:5933–5947

Zhou S, Liu R, Baroudy BM, Malcolm BA, Reyes GR (2003) The effect of ribavirin and IMPDH inhibitors on hepatitis C virus subgenomic replicon RNA. Virology 310:333–342

Subject Index

Current Topics in Microbiology and Immunology

Volumes published since 1989 (and still available)

Vol. 244: **Daëron, Marc; Vivier, Eric (Eds.):** Immunoreceptor Tyrosine-Based Inhibition Motifs. 1999. 20 figs. VIII, 179 pp. ISBN 3-540-65789-4

Vol. 245/I: **Justement, Louis B.; Siminovitch, Katherine A. (Eds.):** Signal Transduction and the Coordination of B Lymphocyte Development and Function I. 2000. 22 figs. XVI, 274 pp. ISBN 3-540-66002-X

Vol. 245/II: **Justement, Louis B.; Siminovitch, Katherine A. (Eds.):** Signal Transduction on the Coordination of B Lymphocyte Development and Function II. 2000. 13 figs. XV, 172 pp. ISBN 3-540-66003-8

Vol. 246: **Melchers, Fritz; Potter, Michael (Eds.):** Mechanisms of B Cell Neoplasia 1998. 1999. 111 figs. XXIX, 415 pp. ISBN 3-540-65759-2

Vol. 247: **Wagner, Hermann (Ed.):** Immunobiology of Bacterial CpG-DNA. 2000. 34 figs. IX, 246 pp. ISBN 3-540-66400-9

Vol. 248: **du Pasquier, Louis; Litman, Gary W. (Eds.):** Origin and Evolution of the Vertebrate Immune System. 2000. 81 figs. IX, 324 pp. ISBN 3-540-66414-9

Vol. 249: **Jones, Peter A.; Vogt, Peter K. (Eds.):** DNA Methylation and Cancer. 2000. 16 figs. IX, 169 pp. ISBN 3-540-66608-7

Vol. 250: **Aktories, Klaus; Wilkins, Tracy, D. (Eds.):** Clostridium difficile. 2000. 20 figs. IX, 143 pp. ISBN 3-540-67291-5

Vol. 251: **Melchers, Fritz (Ed.):** Lymphoid Organogenesis. 2000. 62 figs. XII, 215 pp. ISBN 3-540-67569-8

Vol. 252: **Potter, Michael; Melchers, Fritz (Eds.):** B1 Lymphocytes in B Cell Neoplasia. 2000. XIII, 326 pp. ISBN 3-540-67567-1

Vol. 253: **Gosztonyi, Georg (Ed.):** The Mechanisms of Neuronal Damage in Virus Infections of the Nervous System. 2001. approx. XVI, 270 pp. ISBN 3-540-67617-1

Vol. 254: **Privalsky, Martin L. (Ed.):** Transcriptional Corepressors. 2001. 25 figs. XIV, 190 pp. ISBN 3-540-67569-8

Vol. 255: **Hirai, Kanji (Ed.):** Marek's Disease. 2001. 22 figs. XII, 294 pp. ISBN 3-540-67798-4

Vol. 256: **Schmaljohn, Connie S.; Nichol, Stuart T. (Eds.):** Hantaviruses. 2001, 24 figs. XI, 196 pp. ISBN 3-540-41045-7

Vol. 257: **van der Goot, Gisou (Ed.):** Pore-Forming Toxins, 2001. 19 figs. IX, 166 pp. ISBN 3-540-41386-3

Vol. 258: **Takada, Kenzo (Ed.):** Epstein-Barr Virus and Human Cancer. 2001. 38 figs. IX, 233 pp. ISBN 3-540-41506-8

Vol. 259: **Hauber, Joachim, Vogt, Peter K. (Eds.):** Nuclear Export of Viral RNAs. 2001. 19 figs. IX, 131 pp. ISBN 3-540-41278-6

Vol. 260: **Burton, Didier R. (Ed.):** Antibodies in Viral Infection. 2001. 51 figs. IX, 309 pp. ISBN 3-540-41611-0

Vol. 261: **Trono, Didier (Ed.):** Lentiviral Vectors. 2002. 32 figs. X, 258 pp. ISBN 3-540-42190-4

Vol. 262: **Oldstone, Michael B.A. (Ed.):** Arenaviruses I. 2002, 30 figs. XVIII, 197 pp. ISBN 3-540-42244-7

Vol. 263: **Oldstone, Michael B. A. (Ed.):** Arenaviruses II. 2002, 49 figs. XVIII, 268 pp. ISBN 3-540-42705-8

Vol. 264/I: **Hacker, Jörg; Kaper, James B. (Eds.):** Pathogenicity Islands and the Evolution of Microbes. 2002. 34 figs. XVIII, 232 pp. ISBN 3-540-42681-7

Vol. 264/II: **Hacker, Jörg; Kaper, James B. (Eds.):** Pathogenicity Islands and the Evolution of Microbes. 2002. 24 figs. XVIII, 228 pp. ISBN 3-540-42682-5

Vol. 265: **Dietzschold, Bernhard; Richt, Jürgen A. (Eds.):** Protective and Pathological Immune Responses in the CNS. 2002. 21 figs. X, 278 pp. ISBN 3-540-42668-X

Vol. 266: **Cooper, Koproski (Eds.):** The Interface Between Innate and Acquired Immunity, 2002, 15 figs. XIV, 116 pp. ISBN 3-540-42894-1

Vol. 267: **Mackenzie, John S.; Barrett, Alan D. T.; Deubel, Vincent (Eds.):** Japanese Encephalitis and West Nile Viruses. 2002. 66 figs. X, 418 pp. ISBN 3-540-42783-X

Vol. 268: **Zwickl, Peter; Baumeister, Wolfgang (Eds.):** The Proteasome-Ubiquitin Protein Degradation Pathway. 2002, 17 figs. X, 213 pp. ISBN 3-540-43096-2

Vol. 269: **Koszinowski, Ulrich H.; Hengel, Hartmut (Eds.):** Viral Proteins Counteracting Host Defenses. 2002, 47 figs. XII, 325 pp. ISBN 3-540-43261-2

Vol. 270: **Beutler, Bruce; Wagner, Hermann (Eds.):** Toll-Like Receptor Family Members and Their Ligands. 2002, 31 figs. X, 192 pp. ISBN 3-540-43560-3

Vol. 271: **Koehler, Theresa M. (Ed.):** Anthrax. 2002, 14 figs. X, 169 pp. ISBN 3-540-43497-6

Vol. 272: **Doerfler, Walter; Böhm, Petra (Eds.):** Adenoviruses: Model and Vectors in Virus-Host Interactions. Virion and Structure, Viral Replication, Host Cell Interactions. 2003, 63 figs., approx. 280 pp. ISBN 3-540-00154-9

Vol. 273: **Doerfler, Walter; Böhm, Petra (Eds.):** Adenoviruses: Model and Vectors in Virus-Host Interactions. Immune System, Oncogenesis, Gene Therapy. 2004, 35 figs., approx. 280 pp. ISBN 3-540-06851-1

Vol. 274: **Workman, Jerry L. (Ed.):** Protein Complexes that Modify Chromatin. 2003, 38 figs., XII, 296 pp. ISBN 3-540-44208-1

Vol. 275: **Fan, Hung (Ed.):** Jaagsiekte Sheep Retrovirus and Lung Cancer. 2003, 63 figs., XII, 252 pp. ISBN 3-540-44096-3

Vol. 276: **Steinkasserer, Alexander (Ed.):** Dendritic Cells and Virus Infection. 2003, 24 figs., X, 296 pp. ISBN 3-540-44290-1

Vol. 277: **Rethwilm, Axel (Ed.):** Foamy Viruses. 2003, 40 figs., X, 214 pp. ISBN 3-540-44388-6

Vol. 278: **Salomon, Daniel R.; Wilson, Carolyn (Eds.):** Xenotransplantation. 2003, 22 figs., IX, 254 pp.ISBN 3-540-00210-3

Vol. 279: **Thomas, George; Sabatini, David; Hall, Michael N. (Eds.):** TOR. 2004, 49 figs., X, 364 pp.ISBN 3-540-00534-X

Vol. 280: **Heber-Katz, Ellen (Ed.):** Regeneration: Stem Cells and Beyond. 2004, 42 figs., XII, 194 pp.ISBN 3-540-02238-4

Vol. 281: **Young, John A. T. (Ed.):** Cellular Factors Involved in Early Steps of Retroviral Replication. 2003, 21 figs., IX, 240 pp. ISBN 3-540-00844-6

Vol. 282: **Stenmark, Harald (Ed.):** Phosphoinositides in Subcellular Targeting and Enzyme Activation. 2003, 20 figs., X, 210 pp. ISBN 3-540-00950-7

Vol. 283: **Kawaoka, Yoshihiro (Ed.):** Biology of Negative Strand RNA Viruses: The Power of Reverse Genetics. 2004, 24 figs., IX, 350 pp. ISBN 3-540-40661-1

Vol. 284: **Harris, David (Ed.):** Mad Cow Disease and Related Spongiform Encephalopathies. 2004, 34 figs., IX, 219 pp. ISBN 3-540-20107-6

Vol. 285: **Marsh, Mark (Ed.):** Membrane Trafficking in Viral Replication. 2004, 19 figs., IX, 259 pp. ISBN 3-540-21430-5

Vol. 286: **Madshus, Inger H. (Ed.):** Signalling from Internalized Growth Factor Receptors. 2004, 19 figs., IX, 187 pp. ISBN 3-540-21038-5

Vol. 287: **Enjuanes, Luis (Ed.):** Coronavirus Replication and Reverse Genetics. 2005, 49 figs., XI, 257 pp. ISBN 3-540-21494-1

MIX
Papier aus verantwortungsvollen Quellen
Paper from responsible sources
FSC® C105338

If you have any concerns about our products,
you can contact us on
ProductSafety@springernature.com

In case Publisher is established outside the EU,
the EU authorized representative is:
Springer Nature Customer Service Center GmbH
Europaplatz 3, 69115 Heidelberg, Germany

Printed by Libri Plureos GmbH
in Hamburg, Germany